REED
PROPERTY &
CONSTRUCTION

EXCLUSIVE
EMPLOYMENT ADVICE

PLASTERING

LEVEL 2

B|A BRITISH ASSOCIATION
C|H OF
CONSTRUCTION HEADS

OXFORD
UNIVERSITY PRESS

OXFORD
UNIVERSITY PRESS

Great Clarendon Street, Oxford, OX2 6DP, United Kingdom

Oxford University Press is a department of the University of Oxford. It furthers the University's objective of excellence in research, scholarship, and education by publishing worldwide. Oxford is a registered trade mark of Oxford University Press in the UK and in certain other countries

British Library Cataloguing in Publication Data
Data available

978-1-40-852699-6

1 3 5 7 9 10 8 6 4 2

MIX
Paper from responsible sources
FSC www.fsc.org FSC® C007785

Paper used in the production of this book is a natural, recyclable product made from wood grown in sustainable forests. The manufacturing process conforms to the environmental regulations of the country of origin.

Typeset by GreenGate Publishing Services, Tonbridge, Kent
Printed in Great Britain by Bell and Bain Ltd., Glasgow.

Acknowledgements

The publishers would like to thank the following for permissions to use their photographs:

Cover: © **Lucidio Studio, Inc./Getty**

© **Adrian Sherratt/Alamy**: 5.27, 6.0; © **Anton Starikov/Alamy**: 7.2; © **Helen Sessions/Alamy**: 4.39; © **incamerastock/Alamy**: 6.16; © **Justin Kase zsixz/Alamy**: 4.34, 6.9; © **PHOVOIR/Alamy**: 6.4; © **Steve Atkins Photography/Alamy**: 4.32; **Alexandr Makarov/Shutterstock, yevgeniy11/Shutterstock**: 5.36; **Arcaid Images/Alamy**: 3.16; **Berni/Shutterstock**: 6.1; **blickwinkel/Alamy**: 3.17; **Catnic**: 4.47, 4.48, 6.12; **Chotewang/Shutterstock**: 3.24; **Courtesy of Helen Waller/Everbuild**: 4.38; **DGLowrie/iStock**: 3.23;

ferrantraite/iStock: 3.30; **Filip Ristevski/Shutterstock**: 4.13; **Fotolia**: 1.5, 1.6, 1.7, 1.8, 1.14, 1.15, 1.16, 2.27; **Gary Ombler/Thinkstock**: 4.54; **Hipped end roof. Chicago_bungalow[1] (Wikipedia)**: 3.19; **ictor/iStock**: 3.26; **iStockphoto**: 1.11; **JasonDoiy/iStock**: 2.28; **katylh/iStock**: 3.31; **KjellBrynildsen/iStock**: 3.20; **Mayovskyy Andrew/Shutterstock**: 5.35; **Monkey Business Images/Shutterstock**: 3.0; **ndoeljindoel/Shutterstock**: 2.0; **Nelson Thornes**: 1.9, 1.10, 1.12, 1.13; **ofbeautifulthings/iStock**: 3.22; **Pavel L Photo and Video/Shutterstock**: 3.27; **pejft/iStock**: 3.29; **Peter Davey/Alamy**: 1.0; **PETER GARDINER/SCIENCE PHOTO LIBRARY**: 1.4; **quaximo/Fotolia**: 4.31; **Racatac**: 7.1; **Reproduced by kind permission of Rebecca Nestingen**: 4.41; **Richard Wilson/Oxford University Press**: 4.0, 4.1, 4.2, 4.3, 4.4, 4.5, 4.6, 4.7, 4.8, 4.9, 4.10, 4.11, 4.12, 4.14, 4.16, 4.17, 4.18, 4.19, 4.20, 4.21, 4.22, 4.23, 4.24, 4.25, 4.26, 4.27, 4.28, 4.29, 4.33, 4.36, 4.37, 4.40, 4.45, 4.46, 4.50, 4.51, 4.52, 4.55, 4.56, 4.58, 4.59, 4.60, 4.61, 4.62, 4.63, 4.64, 4.66, 4.78, 4.79, 4.80, 4.81, 4.82, 4.83, 4.84, 4.85, 4.86, 4.87, 4.88, 4.89, 4.90, 4.91, 4.92, 4.93, 4.94, 4.95, 4.96, 4.97, 4.98, 4.99, 4.100, 4.101, 4.102, 4.103, 4.104, 4.105, 4.106, 4.107, 4.108, 4.109, 4.110, 4.111, 4.112, 4.113, 4.114, 4.115, 4.116, 4.117, 4.118, 4.119, 4.120, 5.0, 5.1, 5.3, 5.4, 5.5, 5.6, 5.8, 5.9, 5.10, 5.11, 5.12, 5.13, 5.14, 5.15, 5.16, 5.17, 5.18, 5.19, 5.20, 5.21, 5.22, 5.23, 5.24, 5.25, 5.26, 5.28, 5.29, 5.30, 5.31, 5.33, 5.38, 5.44, 5.45, 5.46, 5.47, 5.48, 5.49, 5.50, 5.51, 5.52, 5.53, 5.54, 5.55, 5.56, 5.57, 5.58, 5.59, 5.60, 5.61, 5.62, 5.63, 5.64, 5.65, 6.7, 6.10, 6.11, 6.13, 6.19, 6.20, 6.21, 6.22, 6.23, 6.24, 6.25, 7.0, 7.3, 7.16, 7.18, 7.23, 7.24, 7.25, 7.26, 7.27, 7.28, 7.29, 7.30, 8.0, 8.1, 8.2, 8.3, 8.4, 8.5, 8.6, 8.7, 8.8, 8.9, 8.10, 8.11, 8.12, 8.14, 8.15, 8.16, 8.17, 8.18, 8.19, 8.20, 8.22, 8.25, 8.26, 8.27, 8.28, 8.34, 8.35, 8.36, 8.37, 8.38, 8.42, 8.43, 8.44, 8.45, 8.46, 8.47, 8.48, 8.49, 8.50, 8.51, 8.52, 8.53, 8.54, 8.55, 8.56, 8.57, 8.58, 8.59, 8.60, 8.61, 8.62, 8.63, 8.64, 8.65, 8.66, 8.67; 8.68; 8.69; 8.70; 8.71; 8.72; 8.73; 8.74; 8.75; 8.76; 8.77; 8.78; 8.79; 8.80; 8.81; 9.0, 9.3; 9.6; 9.8; 9.13; 9.14; 9.15; 9.16; 9.17; 9.18; 9.19; 9.20; 9.21; 9.22; **richsouthwales/Shutterstock**: 3.18; **Ridofranz/iStock**: 4.30; **S. Kuelcue/Shutterstock**: 6.14; **small_frog/iStock**: 3.7; **Stocksnapper/Shutterstock**: 8.39; **stocksolutions/Shutterstock**: 9.1; **Susan Law Cain/Shutterstock**: 3.15; **TanArt/Shutterstock**: 6.5; **titine974/iStock**: 3.25; **TTS Tooltechnic Systems GB Ltd.**: 4.15; **vast natalia/Alamy**: 7.17; **vesilvio/iStock**: 6.2; **WDG Photo/Shutterstock**: 6.15; **Willow Creek Paving Stones**: 7.4; **worldthroughthelens-DIY/Alamy**: 7.11; **www.contenti.com**: 8.21; **www.tanks-direct.co.uk**: 8.13; **YK/Shutterstock**: 3.28; **Yuliya Evstratenko/Shutterstock**: 5.32

Note to learners and tutors

This book clearly states that a risk assessment should be undertaken and the correct PPE worn for the particular activities before any practical activity is carried out. Risk assessments were carried out before photographs for this book were taken and the models are wearing the PPE deemed appropriate for the activity and situation. This was correct at the time of going to print. Colleges may prefer that their learners wear additional items of PPE not featured in the photographs in this book and should instruct learners to do so in the standard risk assessments they hold for activities undertaken by their learners. Learners should follow the standard risk assessments provided by their college for each activity they undertake which will determine the PPE they wear.

CONTENTS

INTRODUCTION

About this book

This book has been written for the Cskills Awards Level 2 Diploma in Plastering. It covers all the units of the qualification, so you can feel confident that your book fully covers the requirements of your course.

This book contains a number of features to help you acquire the knowledge you need. It also demonstrates the practical skills you will need to master to successfully complete your qualification. We've included additional features to show how the skills and knowledge can be applied to the workplace, as well as tips and advice on how you can improve your chances of gaining employment.

The features include:

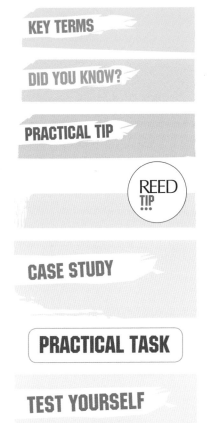

* chapter openers which list the learning outcomes you must achieve in each unit

* key terms that provide explanations of important terminology that you will need to know and understand

* Did you know? margin notes to provide key facts that are helpful to your learning

* practical tips to explain facts or skills to remember when undertaking practical tasks

* Reed tips to offer advice about work, building your CV and how to apply the skills and knowledge you have learnt in the workplace

* case studies that are based on real tradespeople who have undertaken apprenticeships and explain why the skills and knowledge you learn with your training provider are useful in the workplace

* practical tasks that provide step-by-step directions and illustrations for a range of projects you may do during your course

* Test yourself multiple choice questions that appear at the end of each unit to give you the chance to revise what you have learnt and to practise your assessment (your tutor will give you the answers to these questions).

Further support for this book can be found at our website, www.planetvocational.com/subjects/build

CONTRIBUTORS TO THIS BOOK

British Association of Construction Heads

The British Association of Construction Heads is an association formed largely from those managing and delivering the construction curriculum from pre-apprenticeship to post graduate level. The Association is a voluntary organisation and was formed in 1983 and has grown to a position where it can demonstrate that BACH members now manage over 90% of the Learners studying the construction curriculum and includes membership of 80% of the Colleges offering the Construction curriculum in England, Northern Ireland, Scotland and Wales. It accepts membership applications from Colleges and other organisations who are passionate about quality and standards in construction education and training. Visit www.bach.uk.com for more information.

A huge thank you to Mike Morson at Riverside College and David Kehoe at Vision West Nottinghamshire College for their technical expertise in reviewing, advising and facilitating the photo shoot.

Reed Property & Construction

Reed Property & Construction specialises in placing staff at all levels, in both temporary and permanent positions, across the complete lifecycle of the construction process. Our consultants work with most major construction companies in the UK and our clients are involved with the design, build and maintenance of infrastructure projects throughout the UK.

Expert help
As a leading recruitment consultancy for mid–senior level construction staff in the UK, Reed Property & Construction is ideally placed to advise new workers entering the sector, from building a CV to providing expertise and sharing our extensive sector knowledge with you. That's why you will find helpful hints from our highly experienced consultants, designed to help you find that first step on the construction career ladder. These tips range from advice on CV writing to interview tips and techniques, and are linked with the learning material in this book.

Work-related advice
Reed Property & Construction has gained insights from some of our biggest clients to help you understand the mind-set of potential employers. This includes the traits and skills that they would like to see in their new employees, why you need the skills taught in this book and how they are used on a day to day basis within their organisations.

Getting your first job
This invaluable information is not available anywhere else and is geared to helping you gain a position with an employer once you've completed your studies. Entry level positions are not usually offered by recruitment companies, but our advice will help you to apply for jobs in construction and hopefully gain your first position as a skilled worker.

CONTRIBUTORS TO THIS BOOK

The case studies in this book feature staff from Laing O'Rourke and South Tyneside Homes.

Laing O'Rourke is an international engineering company that constructs large-scale building projects all over the world. Originally formed from two companies, John Laing (founded in 1848) and R O'Rourke and Son (founded in 1978) joined forces in 2001.

At Laing O'Rourke, there is a strong and unique apprenticeship programme. It runs a four-year 'Apprenticeship Plus' scheme in the UK, combining formal college education with on-the-job training. Apprentices receive support and advice from mentors and experienced tradespeople, and are given the option of three different career pathways upon completion: remaining on site, continuing into a further education programme, or progressing into supervision and management.

The company prides itself on its people development, supporting educational initiatives and investing in its employees. Laing O'Rourke believes in collaboration and teamwork as a path to achieving greater success, and strives to maintain exceptionally high standards in workplace health and safety.

South Tyneside Council's
Housing Company

South Tyneside Homes was launched in 2006, and was previously part of South Tyneside Council. It now works in partnership with the council to repair and maintain 18,000 properties within the borough, including delivering parts of the Decent Homes Programme.

South Tyneside Homes believes in putting back into the community, with 90 per cent of its employees living in the borough itself. Equality and diversity, as well as health and wellbeing of staff, is a top priority, and it has achieved the Gold Status Investors in People Award.

South Tyneside Homes is committed to the development of its employees, providing opportunities for further education and training and great career paths within the company – 80 per cent of its management team started as apprentices with the company. As well as looking after its staff and their community, the company looks after the environment too, running a renewable energy scheme for council tenants in order to reduce carbon emissions and save tenants money.

The apprenticeship programme at South Tyneside Homes has been recognised nationally, having trained over 80 young people in five main trade areas over the past six years. One of the UK's Top 100 Apprenticeship Employers, it is an Ambassador on the panel of the National Apprentice Service. It has won the Large Employer of the Year Award at the National Apprenticeship Awards and several of its apprentices have been nominated for awards, including winning the Female Apprentice of the Year for the local authority.

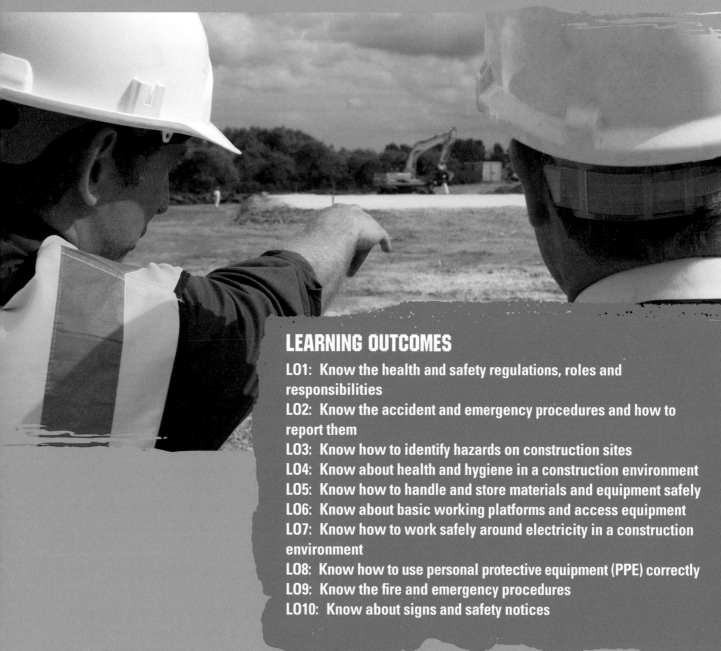

Chapter 1

Unit CSA–L1Core01

HEALTH, SAFETY AND WELFARE IN CONSTRUCTION AND ASSOCIATED INDUSTRIES

LEARNING OUTCOMES

LO1: Know the health and safety regulations, roles and responsibilities

LO2: Know the accident and emergency procedures and how to report them

LO3: Know how to identify hazards on construction sites

LO4: Know about health and hygiene in a construction environment

LO5: Know how to handle and store materials and equipment safely

LO6: Know about basic working platforms and access equipment

LO7: Know how to work safely around electricity in a construction environment

LO8: Know how to use personal protective equipment (PPE) correctly

LO9: Know the fire and emergency procedures

LO10: Know about signs and safety notices

INTRODUCTION

The aim of this chapter is to:

* help you to source relevant safety information
* help you to use the relevant safety procedures at work.

KEY TERMS

HASAWA

– the Health and Safety at Work etc. Act outlines your and your employer's health and safety responsibilities.

COSHH

– the Control of Substances Hazardous to Health Regulations are concerned with controlling exposure to hazardous materials.

DID YOU KNOW?

In 2011 to 2012, there were 49 fatal accidents in the construction industry in the UK. (*Source* HSE, www.hse.gov.uk)

KEY TERMS

HSE

– the Health and Safety Executive, which ensures that health and safety laws are followed.

Accident book

– this is required by law under the Social Security (Claims and Payments) Regulations 1979. Even minor accidents need to be recorded by the employer. For the purposes of RIDDOR, hard copy accident books or online records of incidents are equally acceptable.

HEALTH AND SAFETY REGULATIONS, ROLES AND RESPONSIBILITIES

The construction industry can be dangerous, so keeping safe and healthy at work is very important. If you are not careful, you could injure yourself in an accident or perhaps use equipment or materials that could damage your health. Keeping safe and healthy will help ensure that you have a long and injury-free career.

Although the construction industry is much safer today than in the past, more than 2,000 people are injured and around 50 are killed on site every year. Many others suffer from long-term ill-health such as deafness, spinal damage, skin conditions or breathing problems.

Key health and safety legislation

Laws have been created in the UK to try to ensure safety at work. Ignoring the rules can mean injury or damage to health. It can also mean losing your job or being taken to court.

The two main laws are the Health and Safety at Work etc. Act **(HASAWA)** and the Control of Substances Hazardous to Health Regulations **(COSHH)**.

The Health and Safety at Work etc. Act (HASAWA) (1974)
This law applies to all working environments and to all types of worker, sub-contractor, employer and all visitors to the workplace. It places a duty on everyone to follow rules in order to ensure health, safety and welfare. Businesses must manage health and safety risks, for example by providing appropriate training and facilities. The Act also covers first aid, accidents and ill health.

Reporting of Injuries, Diseases and Dangerous Occurrences Regulations (RIDDOR) (1995)
Under RIDDOR, employers are required to report any injuries, diseases or dangerous occurrences to the **Health and Safety Executive (HSE)**. The regulations also state the need to maintain an **accident book**.

Control of Substances Hazardous to Health (COSHH) (2002)

In construction, it is common to be exposed to substances that could cause ill health. For example, you may use oil-based paints or preservatives, or work in conditions where there is dust or bacteria.

Employers need to protect their employees from the risks associated with using hazardous substances. This means assessing the risks and deciding on the necessary precautions to take.

Any control measures (things that are being done to reduce the risk of people being hurt or becoming ill) have to be introduced into the workplace and maintained; this includes monitoring an employee's exposure to harmful substances. The employer will need to carry out health checks and ensure that employees are made aware of the dangers and are supervised.

Control of Asbestos at Work Regulations (2012)

Asbestos was a popular building material in the past because it was a good insulator, had good fire protection properties and also protected metals against corrosion. Any building that was constructed before 2000 is likely to have some asbestos. It can be found in pipe insulation, boilers and ceiling tiles. There is also asbestos cement roof sheeting and there is a small amount of asbestos in decorative coatings such as Artex.

Asbestos has been linked with lung cancer, other damage to the lungs and breathing problems. The regulations require you and your employer to take care when dealing with asbestos:

* You should always assume that materials contain asbestos unless it is obvious that they do not.

* A record of the location and condition of asbestos should be kept.

* A risk assessment should be carried out if there is a chance that anyone will be exposed to asbestos.

The general advice is as follows:

* Do not remove the asbestos. It is not a hazard unless it is removed or damaged.

* Remember that not all asbestos presents the same risk. Asbestos cement is less dangerous than pipe insulation.

* Call in a specialist if you are uncertain.

Provision and Use of Work Equipment Regulations (PUWER) (1998)

PUWER concerns health and safety risks related to equipment used at work. It states that any risks arising from the use of equipment must either be prevented or controlled, and all suitable safety measures must have been taken. In addition, tools need to be:

* suitable for their intended use

* safe

REED TIP
...
Employers will want to know that you understand the importance of health and safety. Make sure you know the reasons for each safe working practice.

* well maintained

* used only by those who have been trained to do so.

Manual Handling Operations Regulations (1992)

These regulations try to control the risk of injury when lifting or handling bulky or heavy equipment and materials. The regulations state as follows:

* Hazardous manual handling should be avoided if possible.

* An assessment of hazardous manual handling should be made to try to find alternatives.

* You should use mechanical assistance where possible.

* The main idea is to look at how manual handling is carried out and finding safer ways of doing it.

Personal Protection at Work Regulations (PPE) (1992)

This law states that employers must provide employees with personal protective equipment (**PPE**) at work whenever there is a risk to health and safety. PPE needs to be:

* suitable for the work being done

* well maintained and replaced if damaged

* properly stored

* correctly used (which means employees need to be trained in how to use the PPE properly).

Work at Height Regulations (2005)

Whenever a person works at any height there is a risk that they could fall and injure themselves. The regulations place a duty on employers or anyone who controls the work of others. This means that they need to:

* plan and organise the work

* make sure those working at height are **competent**

* assess the risks and provide appropriate equipment

* manage work near or on fragile surfaces

* ensure equipment is inspected and maintained.

In all cases the regulations suggest that, if it is possible, work at height should be avoided. Perhaps the job could be done from ground level? If it is not possible, then equipment and other measures are needed to prevent the risk of falling. When working at height measures also need to be put in place to minimise the distance someone might fall.

KEY TERMS

PPE

– personal protective equipment can include gloves, goggles and hard hats.

Competent

– to be competent an organisation or individual must have:

* sufficient knowledge of the tasks to be undertaken and the risks involved

* the experience and ability to carry out their duties in relation to the project, to recognise their limitations and take appropriate action to prevent harm to those carrying out construction work, or those affected by the work.

(*Source* HSE)

Figure 1.1 Examples of personal protective equipment

Employer responsibilities under HASAWA

HASAWA states that employers with five or more staff need their own health and safety policy. Employers must assess any risks that may be involved in their workplace and then introduce controls to reduce these risks. These risk assessments need to be reviewed regularly.

Employers also need to supply personal protective equipment (PPE) to all employees when it is needed and to ensure that it is worn when required.

Specific employer responsibilities are outlined in Table 1.1.

Employee responsibilities under HASAWA

HASAWA states that all those operating in the workplace must aim to work in a safe way. For example, they must wear any PPE provided and look after their equipment. Employees should not be charged for PPE or any actions that the employer needs to take to ensure safety.

Specific employer responsibilities are outlined in Table 1.1. Table 1.2 identifies the key employee responsibilities.

KEY TERMS

Risk

– the likelihood that a person may be harmed if they are exposed to a hazard.

Hazard

– a potential source of harm, injury or ill-health.

Near miss

– any incident, accident or emergency that did not result in an injury but could have done so.

Employer responsibility	Explanation
Safe working environment	Where possible all potential risks and hazards should be eliminated.
Adequate staff training	When new employees begin a job their induction should cover health and safety. There should be ongoing training for existing employees on risks and control measures.
Health and safety information	Relevant information related to health and safety should be available for employees to read and have their own copies.
Risk assessment	Each task or job should be investigated and potential risks identified so that measures can be put in place. A risk assessment and method statement should be produced. The method statement will tell you how to carry out the task, what PPE to wear, equipment to use and the sequence of its use.
Supervision	A competent and experienced individual should always be available to help ensure that health and safety problems are avoided.

Table 1.1 Employer responsibilities under HASAWA

Employee responsibility	Explanation
Working safely	Employees should take care of themselves, only do work that they are competent to carry out and remove obvious hazards if they are seen.
Working in partnership with the employer	Co-operation is important and you should never interfere with or misuse any health and safety signs or equipment. You should always follow the site rules.
Reporting hazards, near misses and accidents correctly	Any health and safety problems should be reported and discussed, particularly a near miss or an actual accident.

Table 1.2 Employee responsibilities under HASAWA

KEY TERMS

Improvement notice

– this is issued by the HSE if a health or safety issue is found and gives the employer a time limit to make changes to improve health and safety.

Prohibition notice

– this is issued by the HSE if a health or safety issue involving the risk of serious personal injury is found and stops all work until the improvements to health and safety have been made.

Sub-contractor

– an individual or group of workers who are directly employed by the main contractor to undertake specific parts of the work.

Health and Safety Executive

The Health and Safety Executive (HSE) is responsible for health, safety and welfare. It carries out spot checks on different workplaces to make sure that the law is being followed.

HSE inspectors have access to all areas of a construction site and can also bring in the police. If they find a problem then they can issue an **improvement notice**. This gives the employer a limited amount of time to put things right.

In serious cases, the HSE can issue a **prohibition notice**. This means all work has to stop until the problem is dealt with. An employer, the employees or **sub-contractors** could be taken to court.

The roles and responsibilities of the HSE are outlined in Table 1.3.

Responsibility	Explanation
Enforcement	It is the HSE's responsibility to reduce work-related death, injury and ill health. It will use the law against those who put others at risk.
Legislation and advice	The HSE will use health and safety legislation to serve improvement or prohibition notices or even to prosecute those who break health and safety rules. Inspectors will provide advice either face-to-face or in writing on health and safety matters.
Inspection	The HSE will look at site conditions, standards and practices and inspect documents to make sure that businesses and individuals are complying with health and safety law.

Table 1.3 HSE roles and responsibilities

Sources of health and safety information

There is a wide variety of health and safety information. Most of it is available free of charge, while other organisations may make a charge to provide information and advice. Table 1.4 outlines the key sources of health and safety information.

Source	Types of information	Website
Health and Safety Executive (HSE)	The HSE is the primary source of work-related health and safety information. It covers all possible topics and industries.	www.hse.gov.uk
Construction Industry Training Board (CITB)	The national training organisation provides key information on legislation and site safety.	www.citb.co.uk
British Standards Institute (BSI)	Provides guidelines for risk management, PPE, fire hazards and many other health and safety-related areas.	www.bsigroup.com
Royal Society for the Prevention of Accidents (RoSPA)	Provides training, consultancy and advice on a wide range of health and safety issues that are aimed to reduce work related accidents and ill health.	www.rospa.com
Royal Society for Public Health (RSPH)	Has a range of qualifications and training programmes focusing on health and safety.	www.rsph.org.uk

Table 1.4 Health and safety information

Informing the HSE

The HSE requires the reporting of:

* deaths and injuries – any **major injury**, **over 7-day injury** or death

* occupational disease

* dangerous occurrence – a collapse, explosion, fire or collision

* gas accidents – any accidental leaks or other incident related to gas.

Enforcing guidance

Work-related injuries and illnesses affect huge numbers of people. According to the HSE, 1.1 million working people in the UK suffered from a work-related illness in 2011 to 2012. Across all industries, 173 workers were killed, 111,000 other injuries were reported and 27 million working days were lost.

The construction industry is a high risk one and, although only around 5 per cent of the working population is in construction, it accounts for 10 per cent of all major injuries and 22 per cent of fatal injuries.

The good news is that enforcing guidance on health and safety has driven down the numbers of injuries and deaths in the industry. Only 20 years ago over 120 construction workers died in workplace accidents each year. This is now reduced to fewer than 60 a year.

However, there is still more work to be done and it is vital that organisations such as the HSE continue to enforce health and safety and continue to reduce risks in the industry.

On-site safety inductions and toolbox talks

The HSE suggests that all new workers arriving on site should attend a short induction session on health and safety. It should:

* show the commitment of the company to health and safety

* explain the health and safety policy

* explain the roles individuals play in the policy

* state that each individual has a legal duty to contribute to safe working

* cover issues like excavations, work at height, electricity and fire risk

* provide a layout of the site and show evacuation routes

* identify where fire fighting equipment is located

* ensure that all employees have evidence of their skills

* stress the importance of signing in and out of the site.

KEY TERMS

Major injury

– any fractures, amputations, dislocations, loss of sight or other severe injury.

Over 7-day injury

– an injury that has kept someone off work for more than seven days.

DID YOU KNOW?

Workplace injuries cost the UK £13.4bn in 2010 to 2011.

Behaviour and actions that could affect others

It is the responsibility of everyone on site not only to look after their own health and safety, but also to ensure that their actions do not put anyone else at risk.

Trying to carry out work that you are not competent to do is not only dangerous to yourself but could compromise the safety of others.

Simple actions, such as ensuring that all of your rubbish and waste is properly disposed of, will go a long way to removing hazards on site that could affect others.

Just as you should not create a hazard, ignoring an obvious one is just as dangerous. You should always obey site rules and particularly the health and safety rules. You should follow any instructions you are given.

ACCIDENT AND EMERGENCY PROCEDURES

All sites will have specific procedures for dealing with accidents and emergencies. An emergency will often mean that the site needs to be evacuated, so you should know in advance where to assemble and who to report to. The site should never be re-entered without authorisation from an individual in charge or the emergency services.

Types of emergencies

Emergencies are incidents that require immediate action. They can include:

* fires
* spillages or leaks of chemicals or other hazardous substances, such as gas
* failure of a scaffold
* collapse of a wall or trench
* a health problem
* an injury
* bombs and security alerts.

Legislation and reporting accidents

RIDDOR (1995) puts a duty on employers, anyone who is self-employed, or an individual in control of the work, to report any serious workplace accidents, occupational diseases or dangerous occurrences (also known as near misses).

Figure 1.2 It's important that you know where your company's fire-fighting equipment is located

The report has to be made by these individuals and, if it is serious enough, the responsible person may have to fill out a RIDDOR report.

Injuries, diseases and dangerous occurrences

Construction sites can be dangerous places, as we have seen. The HSE maintains a list of all possible injuries, diseases and dangerous occurrences, particularly those that need to be reported.

Injuries

There are two main classifications of injuries: minor and major. A minor injury can usually be handled by a competent first aider, although it is often a good idea to refer the individual to their doctor or to the hospital. Typical minor injuries can include:

* minor cuts   * minor burns   * exposure to fumes.

Major injuries are more dangerous and will usually require the presence of an ambulance with paramedics. Major injuries can include:

* bone fracture   * concussion

* unconsciousness   * electric shock.

Diseases

There are several different diseases and health issues that have to be reported, particularly if a doctor notifies that a disease has been diagnosed. These include:

* poisoning   * infections

* skin diseases   * occupational cancer

* lung diseases   * hand/arm vibration syndrome.

Dangerous occurrences

Even if something happens that does not result in an injury, but could easily have done so, it is classed as a dangerous occurrence. It needs to be reported immediately and then followed up by an accident report form. Dangerous occurrences can include:

* accidental release of a substance that could damage health

* anything coming into contact with overhead power lines

* an electrical problem that caused a fire or explosion

* collapse or partial collapse of scaffolding over 5 m high.

PRACTICAL TIP

An up-to-date list of dangerous occurrences is maintained by the Health and Safety Executive.

Recording accidents and emergencies

The Reporting of Injuries, Diseases and Dangerous Occurrences Regulations (RIDDOR) (1995) requires employers to:

* report any relevant injuries, diseases or dangerous occurrences to the Health and Safety Executive (HSE)

* keep records of incidents in a formal and organised manner (for example, in an accident book or online database).

After an accident, you may need to complete an accident report form – either in writing or online. This form may be completed by the person who was injured or the first aider.

On the accident report form you need to note down:

* the casualty's personal details, e.g. name, address, occupation
* the name of the person filling in the report form
* the details of the accident.

In addition, the person reporting the accident will need to sign the form.

On site a trained first aider will be the first individual to try and deal with the situation. In addition to trying to save life, stop the condition from getting worse and getting help, they will also record the occurrence.

On larger sites there will be a health and safety officer, who would keep records and documentation detailing any accidents and emergencies that have taken place on site. All companies should keep such records; it may be a legal requirement for them to do so under RIDDOR and it is good practice to do so in case the HSE asks to see it.

Importance of reporting accidents and near misses

Reporting incidents is not just about complying with the law or providing information for statistics. Each time an accident or near miss takes place it means lessons can be learned and future problems avoided.

The accident or near miss can alert the business or organisation to a potential problem. They can then take steps to ensure that it does not occur in the future.

Major and minor injuries and near misses

RIDDOR defines a major injury as:

* a fracture (but not to a finger, thumb or toes)
* a dislocation
* an amputation
* a loss of sight in an eye
* a chemical or hot metal burn to the eye
* a penetrating injury to the eye
* an electric shock or electric burn leading to unconsciousness and/or requiring resuscitation
* hyperthermia, heat-induced illness or unconsciousness
* asphyxia
* exposure to a harmful substance
* inhalation of a substance
* acute illness after exposure to toxins or infected materials.

A minor injury could be considered as any occurrence that does not fall into any of the above categories.

A near miss is any incident that did not actually result in an injury but which could have caused a major injury if it had done so. Non-reportable near misses are useful to record as they can help to identify potential problems. Looking at a list of near misses might show patterns for potential risk.

Accident trends

We have already seen that the HSE maintains statistics on the number and types of construction accidents. The following are among the 2011/2012 construction statistics:

* There were 49 fatalities.

* There were 5,000 occupational cancer patients.

* There were 74,000 cases of work-related ill health.

* The most common types of injury were caused by falls, although many injuries were caused by falling objects, collapses and electricity. A number of construction workers were also hurt when they slipped or tripped, or were injured while lifting heavy objects.

Accidents, emergencies and the employer

Even less serious accidents and injuries can cost a business a great deal of money. But there are other costs too:

* Poor company image – if a business does not have health and safety controls in place then it may get a reputation for not caring about its employees. The number of accidents and injuries may be far higher than average.

* Loss of production – the injured individual might have to be treated and then may need a period of time off work to recover. The loss of production can include those who have to take time out from working to help the injured person and the time of a manager or supervisor who has to deal with all the paperwork and problems.

* Insurance – each time there is an accident or injury claim against the company's insurance the premiums will go up. If there are many accidents and injuries the business may find it impossible to get insurance. It is a legal requirement for a business to have insurance so in the end that company might have to close down.

* Closure of site – if there is a serious accident or injury then the site may have to be closed while investigations take place to discover the reason, or who was responsible. This could cause serious delays and loss of income for workers and the business.

DID YOU KNOW?

RoSPA (the Royal Society for the Prevention of Accidents) uses many of the statistics from the HSE. The latest figures that RoSPA has analysed date back to 2008/2009. In that year, 1.2 million people in the UK were suffering from work-related illnesses. With fewer than 132,000 reportable injuries at work, this is believed to be around half of the real figure.

DID YOU KNOW?

An employee working in a small business broke two bones in his arm. He could not return to proper duties for eight months. He lost out on wages while he was off sick and, in total, it cost the business over £45,000.

REED TIP

On some construction sites, you may get a Health and Safety Inspector come to look round without any notice – one more reason to always be thinking about working safely.

Accident and emergency authorised personnel

Several different groups of people could be involved in dealing with accident and emergency situations. These are listed in Table 1.5.

Authorised personnel	Role
First aiders and emergency responders	These are employees on site and in the workforce who have been trained to be the first to respond to accidents and injuries. The minimum provision of an appointed person would be someone who has had basic first aid training. The appointment of a first aider is someone who has attained a higher or specific level of training. A construction site with fewer than 5 employees needs an appointed first aider. A construction site with up to 50 employees requires a trained first aider, and for bigger sites at least one trained first aider is required for every 50 people.
Supervisors and managers	These have the responsibility of managing the site and would have to organise the response and contact emergency services if necessary. They would also ensure that records of any accidents are completed and up to date and notify the HSE if required.
Health and Safety Executive	The HSE requires businesses to investigate all accidents and emergencies. The HSE may send an inspector, or even a team, to investigate and take action if the law has been broken.
Emergency services	Calling the emergency services depends on the seriousness of the accident. Paramedics will take charge of the situation if there is a serious injury and if they feel it necessary will take the individual to hospital.

Table 1.5 People who deal with accident and emergency situations

DID YOU KNOW?

The three main emergency services in the UK are: the Fire Service (for fire and rescue); the Ambulance Service (for medical emergencies); the Police (for an immediate police response). Call them on 999 only if it is an emergency.

The basic first aid kit

BS 8599 relates to first aid kits, but it is not legally binding. The contents of a first aid box will depend on an employer's assessment of their likely needs. The HSE does not have to approve the contents of a first aid box but it states that where the work involves low level hazards the minimum contents of a first aid box should be:

* a copy of its leaflet on first aid – *HSE Basic advice on first aid at work*

* 20 sterile plasters of assorted size

* 2 sterile eye pads

* 4 sterile triangular bandages

* 6 safety pins

* 2 large sterile, unmedicated wound dressings

* 6 medium-sized sterile unmedicated wound dressings

* 1 pair of disposable gloves.

The HSE also recommends that no tablets or medicines are kept in the first aid box.

Figure 1.3 A typical first aid box

What to do if you discover an accident

When an accident happens it may not only injure the person involved directly, but it may also create a hazard that could then injure others. You need to make sure that the area is safe enough for you or someone else to help the injured person. It may be necessary to turn off the electrical supply or remove obstructions to the site of the accident.

The first thing that needs to be done if there is an accident is to raise the alarm. This could mean:

* calling for the first aider

* phoning for the emergency services

* dealing with the problem yourself.

How you respond will depend on the severity of the injury.

You should follow this procedure if you need to contact the emergency services:

* Find a telephone away from the emergency.

* Dial 999.

* You may have to go through a switchboard. Carefully listen to what the operator is saying to you and try to stay calm.

* When asked, give the operator your name and location, and the name of the emergency service or services you require.

* You will then be transferred to the appropriate emergency service, who will ask you questions about the accident and its location. Answer the questions in a clear and calm way.

* Once the call is over, make sure someone is available to help direct the emergency services to the location of the accident.

IDENTIFYING HAZARDS

As we have already seen, construction sites are potentially dangerous places. The most effective way of handling health and safety on a construction site is to spot the hazards and deal with them before they can cause an accident or an injury. This begins with basic housekeeping and carrying out risk assessments. It also means having a procedure in place to report hazards so that they can be dealt with.

Good housekeeping

Work areas should always be clean and tidy. Sites that are messy, strewn with materials, equipment, wires and other hazards can prove to be very dangerous. You should:

* always work in a tidy way

* never block fire exits or emergency escape routes

* never leave nails and screws scattered around

* ensure you clean and sweep up at the end of each working day

* not block walkways

* never overfill skips or bins

* never leave food waste on site.

Risk assessments and method statements

It is a legal requirement for employers to carry out risk assessments. This covers not only those who are actually working on a particular job, but other workers in the immediate area, and others who might be affected by the work.

It is important to remember that when you are carrying out work your actions may affect the safety of other people. It is important, therefore, to know whether there are any potential hazards. Once you know what these hazards are you can do something to either prevent or reduce them as a risk. Every job has potential hazards.

There are five simple steps to carrying out a risk assessment, which are shown in Table 1.6, using the example of repointing brickwork on the front face of a dwelling.

Step	Action	Example
1	Identify hazards	The property is on a street with a narrow pavement. The damaged brickwork and loose mortar need to be removed and placed in a skip below. Scaffolding has been erected. The road is not closed to traffic.
2	Identify who is at risk	The workers repointing are at risk as they are working at height. Pedestrians and vehicles passing are at risk from the positioning of the skip and the chance that debris could fall from height.
3	What is the risk from the hazard that may cause an accident?	The risk to the workers is relatively low as they have PPE and the scaffolding has been correctly erected. The risk to those passing by is higher, as they are unaware of the work being carried out above them.
4	Measures to be taken to reduce the risk	Station someone near the skip to direct pedestrians and vehicles away from the skip while the work is being carried out. Fix a secure barrier to the edge of the scaffolding to reduce the chance of debris falling down. Lower the bricks and mortar debris using a bucket or bag into the skip and not throwing them from the scaffolding. Consider carrying out the work when there are fewer pedestrians and less traffic on the road.
5	Monitor the risk	If there are problems with the first stages of the job, you need to take steps to solve them. If necessary consider taking the debris by hand through the building after removal.

Table 1.6 A five-step risk assessment for repointing brickwork

Your employer should follow these working practices, which can help to prevent accidents or dangerous situations occurring in the workplace:

* *Risk assessments* look carefully at what could cause an individual harm and how to prevent this. This is to ensure that no one should be injured or become ill as a result of their work. Risk assessments identify how likely it is that an accident might happen and the consequences of it happening. A risk factor is worked out and control measures created to try to offset them.

* *Method statements,* however brief, should be available for every risk assessment. They summarise risk assessments and other findings to provide guidance on how the work should be carried out.

* *Permit to work systems* are used for very high risk or even potentially fatal activities. They are checklists that need to be completed before the work begins. They must be signed by a supervisor.

* A *hazard book* lists standard tasks and identifies common hazards. These are useful tools to help quickly identify hazards related to particular tasks.

Types of hazards

Typical construction accidents can include:

* fires and explosions

* slips, trips and falls.

* burns, including those from chemicals

* falls from scaffolding, ladders and roofs

* electrocution

* injury from faulty machinery

* power tool accidents

* being hit by construction debris

* falling through holes in flooring

We will look at some of the more common hazards in a little more detail.

Fires

Fires need oxygen, heat and fuel to burn. Even a spark can provide enough heat needed to start a fire, and anything flammable, such as petrol, paper or wood, provides the fuel. It may help to remember the 'triangle of fire' – heat, oxygen and fuel are all needed to make fire so remove one or more to help prevent or stop the fire.

Tripping

Leaving equipment and materials lying around can cause accidents, as can trailing cables and spilt water or oil. Some of these materials are also potential fire hazards.

Chemical spills

If the chemicals are not hazardous then they just need to be mopped up. But sometimes they do involve hazardous materials and there will be an existing plan on how to deal with them. A risk assessment will have been carried out.

Falls from height

A fall even from a low height can cause serious injuries. Precautions need to be taken when working at height to avoid permanent injury. You should also consider falls into open excavations as falls from height. All the same precautions need to be in place to prevent a fall.

Burns

Burns can be caused not only by fires and heat, but also from chemicals and solvents. Electricity and wet concrete and cement can also burn skin. PPE is often the best way to avoid these dangers. Sunburn is a common and uncomfortable form of burning and sunscreen should be made available. For example, keeping skin covered up will help to prevent sunburn. You might think a tan looks good, but it could lead to skin cancer.

Electrical

Electricity is hazardous and electric shocks can cause burns and muscle damage, and can kill.

Exposure to hazardous substances

We look at hazardous substances in more detail on pages 20–1. COSHH regulations identify hazardous substances and require them to be labelled. You should always follow the instructions when using them.

Plant and vehicles

On busy sites there is always a danger from moving vehicles and heavy plant. Although many are fitted with reversing alarms, it may not be easy to hear them over other machinery and equipment. You should always ensure you are not blocking routes or exits. Designated walkways separate site traffic and pedestrians – this includes workers who are walking around the site. Crossing points should be in place for ease of movement on site.

Reporting hazards

We have already seen that hazards have the potential to cause serious accidents and injuries. It is therefore important to report hazards and there are different methods of doing this.

The first major reason to report hazards is to prevent danger to others, whether they are other employees or visitors to the site. It is vital to prevent accidents from taking place and to quickly correct any dangerous situations.

Injuries, diseases and actual accidents all need to be reported and so do dangerous occurrences. These are incidents that do not result in an actual injury, but could easily have hurt someone.

Accidents need to be recorded in an accident book, computer database or other secure recording system, as do near misses. Again it is a legal requirement to keep appropriate records of accidents and every company will have a procedure for this which they should tell you about. Everyone should know where the book is kept or how the records are made. Anyone that has been hurt or has taken part in dealing with an occurrence should complete the details of what has happened. Typically this will require you to fill in:

* the date, time and place of the incident

* how it happened

* what was the cause

* how it was dealt with

* who was involved

* signature and date.

The details in the book have to be transferred onto an official HSE report form.

As far as is possible, the site, company or workplace will have set procedures in place for reporting hazards and accidents. These procedures will usually be found in the place where the accident book or records are stored. The location tends to be posted on the site notice board.

How hazards are created

Construction sites are busy places. There are constantly new stages in development. As each stage is begun a whole new set of potential hazards need to be considered.

At the same time, new workers will always be joining the site. It is mandatory for them to be given health and safety instruction during induction. But sometimes this is impossible due to pressure of work or availability of trainers.

Construction sites can become even more hazardous in times of extreme weather:

* Flooding – long periods of rain can cause trenches to fill with water, cellars to be flooded and smooth surfaces to become extremely wet and slippery.

* Wind – strong winds may prevent all work at height. Scaffolding may have become unstable, unsecured roofing materials may come loose, dry-stored materials such as sand and cement may have been blown across the site.

* Heat – this can change the behaviour of materials: setting quicker, failing to cure and melting. It can also seriously affect the health of the workforce through dehydration and heat exhaustion.

* Snow – this can add enormous weight to roofs and other structures and could cause collapse. Snow can also prevent access or block exits and can mean that simple and routine work becomes impossible due to frozen conditions.

Storing combustibles and chemicals

A combustible substance can be both flammable and explosive. There are some basic suggestions from the HSE about storing these:

* Ventilation – the area should be well ventilated to disperse any vapours that could trigger off an explosion.

* Ignition – an ignition is any spark or flame that could trigger off the vapours, so materials should be stored away from any area that uses electrical equipment or any tool that heats up.

* Containment – the materials should always be kept in proper containers with lids and there should be spillage trays to prevent any leak seeping into other parts of the site.

* Exchange – in many cases it can be possible to find an alternative material that is less dangerous. This option should be taken if possible.

* Separation – always keep flammable substances away from general work areas. If possible they should be partitioned off.

Combustible materials can include a large number of commonly used substances, such as cleaning agents, paints and adhesives.

DID YOU KNOW?

You do not have to be involved in specialist work to come into contact with combustibles.

HEALTH AND HYGIENE

Just as hazards can be a major problem on site, other less obvious problems relating to health and hygiene can also be an issue. It is both your responsibility and that of your employer to make sure that you stay healthy.

The employer will need to provide basic welfare facilities, no matter where you are working and these must have minimum standards.

KEY TERMS

Contamination

– this is when a substance has been polluted by some harmful substance or chemical.

Welfare facilities

Welfare facilities can include a wide range of different considerations, as can be seen in Table 1.7.

Facilities	Purpose and minimum standards
Toilets	If there is a lock on the door there is no need to have separate male and female toilets. There should be enough for the site workforce. If there is no flushing water on site they must be chemical toilets.
Washing facilities	There should be a wash basin large enough to be able to wash up to the elbow. There should be soap, hot and cold water and, if you are working with dangerous substances, then showers are needed.
Drinking water	Clean drinking water should be available; either directly connected to the mains or bottled water. Employers must ensure that there is no contamination.
Dry room	This can operate also as a store room, which needs to be secure so that workers can leave their belongings there and also use it as a place to dry out if they have been working in wet weather, in which case a heater needs to be provided.
Work break area	This is a shelter out of the wind and rain, with a kettle, a microwave, tables and chairs. It should also have heating.

Table 1.7 Welfare facilities in the workplace

CASE STUDY

South
Tyneside Homes

South Tyneside Council's
Housing Company

Staying safe on site

Johnny McErlane finished his apprenticeship at South Tyneside Homes a year ago.

'I've been working on sheltered accommodation for the last year, so there are a lot of vulnerable and elderly people around. All the things I learnt at college from doing the health and safety exams comes into practice really, like taking care when using extension leads, wearing high-vis and correct footwear. It's not just about your health and safety, but looking out for others as well.

On the shelters, you can get a health and safety inspector who just comes around randomly, so you have to always be ready. It just becomes a habit once it's been drilled into you. You're health and safety conscious all the time.

The shelters also have a fire alarm drill every second Monday, so you've got to know the procedure involved there. When it comes to the more specialised skills, such as mouth-to-mouth and CPR, you might have a designated first aider on site who will have their skills refreshed regularly. Having a full first aid certificate would be valuable if you're working in construction.

You cover quite a bit of the first aid skills in college and you really have to know them because you're not always working on large sites. For example, you might be on the repairs team, working in people's houses where you wouldn't have a first aider, so you've got to have the basic knowledge yourself, just in case. All our vans have a basic first aid kit that's kept fully stocked.

The company keeps our knowledge current with these "toolbox talks", which are like refresher courses. They give you any new information that needs to be passed on to all the trades. It's a good way of keeping everyone up to date.'

Noise

Ear defenders are the best precaution to protect the ears from loud noises on site. Ear defenders are either basic ear plugs or ear muffs, which can be seen in Fig 1.13 on page 32.

The long-term impact of noise depends on the intensity and duration of the noise. Basically, the louder and longer the noise exposure, the more damage is caused. There are ways of dealing with this:

* Remove the source of the noise.

* Move the equipment away from those not directly working with it.

* Put the source of the noise into a soundproof area or cover it with soundproof material.

* Ask a supervisor if they can move all other employees away from that part of the site until the noise stops.

Substances hazardous to health

COSHH Regulations (see page 3) identify a wide variety of substances and materials that must be labelled in different ways.

Controlling the use of these substances is always difficult. Ideally, their use should be eliminated (stopped) or they should be replaced with something less harmful. Failing this, they should only be used in controlled or restricted areas. If none of this is possible then they should only be used in controlled situations.

If a hazardous situation occurs at work, then you should:

* ensure the area is made safe

* inform the supervisor, site manager, safety officer or other nominated person.

You will also need to report any potential hazards or near misses.

Personal hygiene

Construction sites can be dirty places to work. Some jobs will expose you to dust, chemicals or substances that can make contact with your skin or may stain your work clothing. It is good practice to wear suitable PPE as a first line of defence as chemicals can penetrate your skin. Whenever you have finished a job you should always wash your hands. This is certainly true before eating lunch or travelling home. It can be good practice to have dedicated work clothing, which should be washed regularly.

Always ensure you wash your hands and face and scrub your nails. This will prevent dirt, chemicals and other substances from contaminating your food and your home.

Make sure that you regularly wash your work clothing and either repair it or replace it if it becomes too worn or stained.

Health risks

The construction industry uses a wide variety of substances that could harm your health. You will also be carrying out work that could be a health risk to you, and you should always be aware that certain activities could cause long-term damage or even kill you if things go wrong. Unfortunately not all health risks are immediately obvious. It is important to make sure that from time to time you have health checks, particularly if you have been using hazardous substances. Table 1.8 outlines some potential health risks in a typical construction site.

KEY TERMS

Dermatitis

– this is an inflammation of the skin. The skin will become red and sore, particularly if you scratch the area. A GP should be consulted.

Leptospirosis

– this is also known as Weil's disease. It is spread by touching soil or water contaminated with the urine of wild animals infected with the leptospira bacteria. Symptoms are usually flu-like but in extreme cases it can cause organ failure.

Health risk	Potential future problems
Dust	The most dangerous potential dust is, of course, asbestos, which **should only be handled by specialists under controlled conditions**. But even brick dust and other fine particles can cause eye injuries, problems with breathing and even cancer.
Chemicals	Inhaling or swallowing dangerous chemicals could cause immediate, long-term damage to lungs and other internal organs. Skin problems include burns or skin can become very inflamed and sore. This is known as dermatitis.
Bacteria	Contact with waste water or soil could lead to a bacterial infection. The germs in the water or dirt could cause infection which will require treatment if they enter the body. The most extreme version is leptospirosis.
Heavy objects	Lifting heavy, bulky or awkward objects can lead to permanent back injuries that could require surgery. Heavy objects can also damage the muscles in all areas of the body.
Noise	Failure to wear ear defenders when you are exposed to loud noises can permanently affect your hearing. This could lead to deafness in the future.
Vibrating tools	Using machines that vibrate can cause a condition known as hand/arm vibration syndrome (HAVS) or vibration white finger, which is caused by injury to nerves and blood vessels. You will feel tingling that could lead to permanent numbness in the fingers and hands, as well as muscle weakness.
Cuts	Any open wound, no matter how small, leaves your body exposed to potential infections. Cuts should always be cleaned and covered, preferably with a waterproof dressing. The blood loss from deep cuts could make you feel faint and weak, which may be dangerous if you are working at height or operating machinery.
Sunlight	Most construction work involves working outside. There is a temptation to take advantage of hot weather and get a tan. But long-term exposure to sunshine means risking skin cancer so you should cover up and apply sun cream.
Head injuries	You should seek medical attention after any bump to the head. Severe head injuries could cause epilepsy, hearing problems, brain damage or death.

Table 1.8 Health risks in construction

HANDLING AND STORING MATERIALS AND EQUIPMENT

On a busy construction site it is often tempting not to even think about the potential dangers of handling equipment and materials. If something needs to be moved or collected you will just pick it up without any thought. It is also tempting just to drop your tools and other equipment when you have finished with them to deal with later. But abandoned equipment and tools can cause hazards both for you and for other people.

Safe lifting

Lifting or handling heavy or bulky items is a major cause of injuries on construction sites. So whenever you are dealing with a heavy load, it is important to carry out a basic risk assessment.

The first thing you need to do is to think about the job to be done and ask:

* Do I need to lift it manually or is there another way of getting the object to where I need it?

Consider any mechanical methods of transporting loads or picking up materials. If there really is no alternative, then ask yourself:

1. Do I need to bend or twist?
2. Does the object need to be lifted or put down from high up?
3. Does the object need to be carried a long way?
4. Does the object need to be pushed or pulled for a long distance?
5. Is the object likely to shift around while it is being moved?

If the answer to any of these questions is 'yes', you may need to adjust the way the task is done to make it safer.

Think about the object itself. Ask:

1. Is it just heavy or is it also bulky and an awkward shape?
2. How easy is it to get a good hand-hold on the object?
3. Is the object a single item or are there parts that might move around and shift the weight?
4. Is the object hot or does it have sharp edges?

Again, if you have answered 'yes' to any of these questions, then you need to take steps to address these issues.

It is also important to think about the working environment and where the lifting and carrying is taking place. Ask yourself:

1. Are the floors stable?

2. Are the surfaces slippery?

3. Will a lack of space restrict my movement?

4. Are there any steps or slopes?

5. What is the lighting like?

Before lifting and moving an object, think about the following:

* Check that your pathway is clear to where the load needs to be taken.

* Look at the product data sheet and assess the weight. If you think the object is too heavy or difficult to move then ask someone to help you. Alternatively, you may need to use a mechanical lifting device.

When you are ready to lift, gently raise the load. Take care to ensure the correct posture – you should have a straight back, with your elbows tucked in, your knees bent and your feet slightly apart.

Once you have picked up the load, move slowly towards your destination. When you get there, make sure that you do not drop the load but carefully place it down.

DID YOU KNOW?

Although many people regard the weight limit for lifting and/or moving heavy or awkward objects to be 20 kg, the HSE does not recommend safe weights. There are many things that will affect the ability of an individual to lift and carry particular objects and the risk that this creates, so manual handling should be avoided altogether where possible.

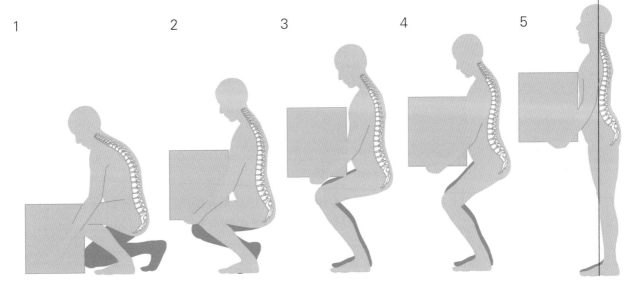

1   2   3   4   5

Figure 1.4 Take care to follow the correct procedure for lifting

Sack trolleys are useful for moving heavy and bulky items around. Gently slide the bottom of the sack trolley under the object and then raise the trolley to an angle of 45° before moving off. Make sure that the object is properly balanced and is not too big for the trolley.

Trailers and forklift trucks are often used on large construction sites, as are dump trucks. Never use these without proper training.

Figure 1.5 Pallet truck   Figure 1.6 Sack trolley

Site safety equipment

You should always read the construction site safety rules and when required wear your PPE. Simple things, such as wearing the right footwear for the right job, are important.

Safety equipment falls into two main categories:

* PPE – including hard hats, footwear, gloves, glasses and safety vests

* perimeter safety – this includes screens, netting and guards or clamps to prevent materials from falling or spreading.

Construction safety is also directed by signs, which will highlight potential hazards.

Safe handling of materials and equipment

All tools and equipment are potentially dangerous. It is up to you to make sure that they do not cause harm to yourself or others. You should always know how to use tools and equipment. This means either instruction from someone else who is experienced, or at least reading the manufacturer's instructions.

You should always make sure that you:

* use the right tool – don't be tempted to use a tool that is close to hand instead of the one that is right for the job

* wear your PPE – the one time you decide not to bother could be the time that you injure yourself

* never try to use a tool or a piece of equipment that you have not been trained to use.

You should always remember that if you are working on a building that was constructed before 2000 it may contain asbestos.

Correct storage

We have already seen that tools and equipment need to be treated with respect. Damaged tools and equipment are not only less effective at doing their job, they could also cause you to injure yourself.

Table 1.9 provides some pointers on how to store and handle different types of materials and equipment.

Materials and equipment	Safe storage and handling
Hand tools	Store hand tools with sharp edges either in a cover or a roll. They should be stored in bags or boxes. They should always be dried before putting them away as they will rust.
Power tools	Never carry them by the cable. Store them in their original carrying case. Always follow the manufacturer's instructions.
Wheelbarrows	Check the tyres and metal stays regularly. Always clean out after use and never overload.
Bricks and blocks	Never store more than two packs high. When cutting open a pack, be careful as the bricks could collapse.
Slabs and curbs	Store slabs flat on their edges on level ground, preferably with wood underneath to prevent damage. Store curbs the same way. To prevent weather damage, cover them with a sheet.
Tiles	Always cover them and protect them from damage as they are relatively fragile. Ideally store them in a hut or container.
Aggregates	Never store aggregates under trees as leaves will drop on them and contaminate them. Cover them with plastic sheets.
Plaster and plasterboard	Plaster needs to be kept dry, so even if stored inside you should take the precaution of putting the bags on pallets. To prevent moisture do not store against walls and do not pile higher than five bags. Plasterboard can be awkward to manage and move around. It also needs to be stored in a waterproof area. It should be stored flat and off the ground but should not be stored against walls as it may bend. Use a rotation system so that the materials are not stored in the same place for long periods.
Wood	Always keep wood in dry, well-ventilated conditions. If it needs to be stored outside it should be stored on bearers that may be on concrete. If wood gets wet and bends it is virtually useless. Always be careful when moving large cuts of wood or sheets of ply or MDF as they can easily become damaged.
Adhesives and paint	Always read the manufacturer's instructions. Ideally they should always be stored on clearly marked shelves. Make sure you rotate the stock using the older stock first. Always make sure that containers are tightly sealed. Storage areas must comply with fire regulations and display signs to advise of their contents.

Table 1.9 Safe storing and handling of materials and equipment

Waste control

The expectation within the building services industry is increasingly that working practices conserve energy and protect the environment. Everyone can play a part in this. For example, you can contribute by turning off hose pipes when you have finished using water, or not running electrical items when you don't need to.

Simple things, such as keeping construction sites neat and orderly, can go a long way to conserving energy and protecting the environment. A good way to remember this is Sort, Set, Shine, Standardise:

* Sort – sort and store items in your work area, eliminate clutter and manage deliveries.

* Set – everything should have its own place and be clearly marked and easy to access. In other words, be neat!

Figure 1.7 It's important to create as little waste as possible on the construction site

* Shine – clean your work area and you will be able to see potential problems far more easily.

* Standardise – by using standardised working practices you can keep organised, clean and safe.

Reducing waste is all about good working practice. By reducing wastage disposal, and recycling materials on site, you will benefit from savings on raw materials and lower transportation costs.

Planning ahead, and accurately measuring and cutting materials, means that you will be able to reduce wastage.

BASIC WORKING PLATFORMS AND ACCESS EQUIPMENT

Working at height should be eliminated or the work carried out using other methods where possible. However, there may be situations where you may need to work at height. These situations can include:

* roofing

* repair and maintenance above ground level

* working on high ceilings.

Any work at height must be carefully planned. Access equipment includes all types of ladder, scaffold and platform. You must always use a working platform that is safe. Sometimes a simple step ladder will be sufficient, but at other times you may have to use a tower scaffold.

Generally, ladders are fine for small, quick jobs of less than 30 minutes. However, for larger, longer jobs a more permanent piece of access equipment will be necessary.

Working platforms and access equipment: good practice and dangers of working at height

Table 1.10 outlines the common types of equipment used to allow you to work at heights, along with the basic safety checks necessary.

Equipment	Main features	Safety checks
Step ladder	Ideal for confined spaces. Four legs give stability	• Knee should remain below top of steps • Check hinges, cords or ropes • Position only to face work
Ladder	Ideal for basic access, short-term work. Made from aluminium, fibreglass or wood	• Check rungs, tie rods, repairs, and ropes and cords on stepladders • Ensure it is placed on firm, level ground • Angle should be no greater than 75° or 1 in 4
Mobile mini towers or scaffolds	These are usually aluminium and foldable, with lockable wheels	• Ensure the ground is even and the wheels are locked • Never move the platform while it has tools, equipment or people on it
Roof ladders and crawling boards	The roof ladder allows access while crawling boards provide a safe passage over tiles	• The ladder needs to be long enough and supported • Check boards are in good condition • Check the welds are intact • Ensure all clips function correctly
Mobile tower scaffolds	These larger versions of mini towers usually have edge protection	• Ensure the ground is even and the wheels are locked • Never move the platform while it has tools, equipment or people on it • Base width to height ratio should be no greater than 1:3
Fixed scaffolds and edge protection	Scaffolds fitted and sized to the specific job, with edge protection and guard rails	• There needs to be sufficient braces, guard rails and scaffold boards • The tubes should be level • There should be proper access using a ladder
Mobile elevated work platforms	Known as scissor lifts or cherry pickers	• Specialist training is required before use • Use guard rails and toe boards • Care needs to be taken to avoid overhead hazards such as cables

Table 1.10 Equipment for working at height and safety checks

You must be trained in the use of certain types of access equipment, like mobile scaffolds. Care needs to be taken when assembling and using access equipment. These are all examples of good practice:

• Step ladders should always rest firmly on the ground. Only use the top step if the ladder is part of a platform.

• Do not rest ladders against fragile surfaces, and always use both hands to climb. It is best if the ladder is steadied (footed) by someone at the foot of the ladder. Always maintain three points of contact – two feet and one hand.

• A roof ladder is positioned by turning it on its wheels and pushing it up the roof. It then hooks over the ridge tiles. Ensure that the access ladder to the roof is directly beside the roof ladder.

• A mobile scaffold is put together by slotting sections until the required height is reached. The working platform needs to have a suitable edge protection such as guard-rails and toe-boards. Always push from the bottom of the base and not from the top to move it, otherwise it may lean or topple over.

Figure 1.8 A tower scaffold

WORKING SAFELY WITH ELECTRICITY

It is essential whenever you work with electricity that you are competent and that you understand the common dangers. Electrical tools must be used in a safe manner on site. There are precautions that you can take to prevent possible injury, or even death.

Precautions

Whether you are using electrical tools or equipment on site, you should always remember the following:

* Use the right tool for the job.

* Use a transformer with equipment that runs on 110V.

* Keep the two voltages separate from each other. You should avoid using 230V where possible but, if you must, use a residual current device (RCD) if you have to use 230V.

* When using 110V, ensure that leads are yellow in colour.

* Check the plug is in good order

* Confirm that the fuse is the correct rating for the equipment.

* Check the cable (including making sure that it does not present a tripping hazard).

* Find out where the mains switch is, in case you need to turn off the power in the event of an emergency.

* Never attempt to repair electrical equipment yourself.

* Disconnect from the mains power before making adjustments, such as changing a drill bit.

* Make sure that the electrical equipment has a sticker that displays a recent test date.

Visual inspection and testing is a three-stage process:

1. The user should check for potential danger signs, such as a frayed cable or cracked plug.

2. A formal visual inspection should then take place. If this is done correctly then most faults can be detected.

3. Combined inspections and **PAT** should take place at regular intervals by a competent person.

Watch out for the following causes of accidents – they would also fail a safety check:

KEY TERMS

PAT

– Portable Appliance Testing – regular testing is a health and safety requirement under the Electricity at Work Regulations (1989).

* damage to the power cable or plug
* taped joints on the cable
* wet or rusty tools and equipment
* weak external casing
* loose parts or screws
* signs of overheating
* the incorrect fuse
* lack of cord grip
* electrical wires attached to incorrect terminals
* bare wires.

When preparing to work on an electrical circuit, do not start until a permit to work has been issued by a supervisor or manager to a competent person.

Make sure the circuit is broken before you begin. A 'dead' circuit will not cause you, or anybody else, harm. These steps must be followed:

* Switch off – ensure the supply to the circuit is switched off by disconnecting the supply cables or using an isolating switch.

* Isolate – disconnect the power cables or use an isolating switch.

* Warn others – to avoid someone reconnecting the circuit, place warning signs at the isolation point.

* Lock off – this step physically prevents others from reconnecting the circuit.

* Testing – is carried out by electricians but you should be aware that it involves three parts:

  1. testing a voltmeter on a known good source (a live circuit) so you know it is working properly

  2. checking that the circuit to be worked on is dead

  3. rechecking your voltmeter on the known live source, to prove that it is still working properly.

It is important to make sure that the correct point of isolation is identified. Isolation can be next to a local isolation device, such as a plug or socket, or a circuit breaker or fuse.

The isolation should be locked off using a unique key or combination. This will prevent access to a main isolator until the work has been completed. Alternatively, the handle can be made detachable in the OFF position so that it can be physically removed once the circuit is switched off.

Dangers

You are likely to encounter a number of potential dangers when working with electricity on construction sites or in private houses. Table 1.11 outlines the most common dangers.

Danger	Identifying the danger
Faulty electrical equipment	Visually inspect for signs of damage. Equipment should be double insulated or incorporate an earth cable.
Damaged or worn cables	Check for signs of wear or damage regularly. This includes checking power tools and any wiring in the property.
Trailing cables	Cables lying on the ground, or worse, stretched too far, can present a tripping hazard. They could also be cut or damaged easily.
Cables and pipe work	Always treat services you find as though they are live. This is very important as services can be mistaken for one another. You may have been trained to use a cable and pipe locator that finds cables and metal pipes.
Buried or hidden cables	Make sure you have plans. Alternatively, use a cable and pipe locator, mark the positions, look out for signs of service connection cables or pipes and hand-dig trial holes to confirm positions.
Inadequate over-current protection	Check circuit breakers and fuses are the correct size current rating for the circuit. A qualified electrician may have to identify and label these.

Table 1.11 Common dangers when working with electricity

Each year there are around 1,000 accidents at work involving electric shocks or burns from electricity. If you are working in a construction site you are part of a group that is most at risk. Electrical accidents happen when you are working close to equipment that you think is disconnected but which is, in fact, live.

Another major danger is when electrical equipment is either misused or is faulty. Electricity can cause fires and contact with the live parts can give you an electric shock or burn you.

Different voltages

The two most common voltages that are used in the UK are 230V and 110V:

* 230V: this is the standard domestic voltage. But on construction sites it is considered to be unsafe and therefore 110V is commonly used.

* 110V: these plugs are marked with a yellow casement and they have a different shaped plug. A transformer is required to convert 230V to 110V.

Some larger homes, as well as industrial and commercial buildings, may have 415V supplies. This is the same voltage that is found on overhead electricity cables. In most houses and other buildings the voltage from these cables is reduced to 230V. This is what most electrical equipment works from. Some larger machinery actually needs 415V.

In these buildings the 415V comes into the building and then can either be used directly or it is reduced so that normal 230V appliances can be used.

Colour coded cables

Normally you will come across three differently coloured wires: Live, Neutral and Earth. These have standard colours that comply with European safety standards and to ensure that they are easily identifiable. However, in some older buildings the colours are different.

Wire type	Modern colour	Older colour
Live	Brown	Red
Neutral	Blue	Black
Earth	Yellow and Green	Yellow and Green

Table 1.12 Colour coding of cables

Working with equipment with different electrical voltages

You should always check that the electrical equipment that you are going to use is suitable for the available electrical supply. The equipment's power requirements are shown on its rating plate. The voltage from the supply needs to match the voltage that is required by the equipment.

Storing electrical equipment

Electrical equipment should be stored in dry and secure conditions. Electrical equipment should never get wet but – if it does happen – it should be dried before storage. You should always clean and adjust the equipment before connecting it to the electricity supply.

PERSONAL PROTECTIVE EQUIPMENT (PPE)

Personal protective equipment, or PPE, is a general term that is used to describe a variety of different types of clothing and equipment that aim to help protect against injuries or accidents. Some PPE you will use on a daily basis and others you may use from time to time. The type of PPE you wear depends on what you are doing and where you are. For example, the practical exercises in this book were photographed at a college, which has rules and requirements for PPE that are different to those on large construction sites. Follow your tutor's or employer's instructions at all times.

Types of PPE

PPE literally covers from head to foot. Here are the main PPE types.

Figure 1.9 A hi-vis jacket

Figure 1.10 Safety glasses and goggles

Figure 1.11 Hand protection

Figure 1.12 Head protection

Figure 1.13 Hearing protection

Protective clothing

Clothing protection such as overalls:

* provides some protection from spills, dust and irritants
* can help protect you from minor cuts and abrasions
* reduces wear to work clothing underneath.

Sometimes you may need waterproof or chemical-resistant overalls.

High visibility (hi-vis) clothing stands out against any background or in any weather conditions. It is important to wear high visibility clothing on a construction site to ensure that people can see you easily. In addition, workers should always try to wear light-coloured clothing underneath, as it is easier to see.

You need to keep your high visibility and protective clothing clean and in good condition.

Employers need to make sure that employees understand the reasons for wearing high visibility clothing and the consequences of not doing so.

Eye protection

For many jobs, it is essential to wear goggles or safety glasses to prevent small objects, such as dust, wood or metal, from getting into the eyes. As goggles tend to steam up, particularly if they are being worn with a mask, safety glasses can often be a good alternative.

Hand protection

Wearing gloves will help to prevent damage or injury to the hands or fingers. For example, general purpose gloves can prevent cuts, and rubber gloves can prevent skin irritation and inflammation, such as contact dermatitis caused by handling hazardous substances. There are many different types of gloves available, including specialist gloves for working with chemicals.

Head protection

Hard hats or safety helmets are compulsory on building sites. They can protect you from falling objects or banging your head. They need to fit well and they should be regularly inspected and checked for cracks. Worn straps mean that the helmet should be replaced, as a blow to the head can be fatal. Hard hats bear a date of manufacture and should be replaced after about 3 years.

Hearing protection

Ear defenders, such as ear protectors or plugs, aim to prevent damage to your hearing or hearing loss when you are working with loud tools or are involved in a very noisy job.

Respiratory protection

Breathing in fibre, dust or some gases could damage the lungs. Dust is a very common danger, so a dust mask, face mask or respirator may be necessary.

Make sure you have the right mask for the job. It needs to fit properly otherwise it will not give you sufficient protection.

Foot protection

Foot protection is compulsory on site, particularly if you are undertaking heavy work. Footwear should include steel toecaps (or equivalent) to protect feet against dropped objects, midsole protection (usually a steel plate) to protect against puncture or penetration from things like nails on the floor and soles with good grip to help prevent slips on wet surfaces.

Figure 1.14 Respiratory protection

Legislation covering PPE

The most important piece of legislation is the Personal Protective Equipment at Work Regulations (1992). It covers all sorts of PPE and sets out your responsibilities and those of the employer. Linked to this are the Control of Substances Hazardous to Health (2002) and the Provision and Use of Work Equipment Regulations (1992 and 1998).

Storing and maintaining PPE

All forms of PPE will be less effective if they are not properly maintained. This may mean examining the PPE and either replacing or cleaning it, or if relevant testing or repairing it. PPE needs to be stored properly so that it is not damaged, contaminated or lost. Each type of PPE should have a CE mark. This shows that it has met the necessary safety requirements.

Importance of PPE

PPE needs to be suitable for its intended use and it needs to be used in the correct way. As a worker or an employee you need to:

* make sure you are trained to use PPE

* follow your employer's instructions when using the PPE and always wear it when you are told to do so

* look after the PPE and if there is a problem with it report it.

Your employer will:

* know the risks that the PPE will either reduce or avoid

* know how the PPE should be maintained

* know its limitations.

Consequences of not using PPE

The consequences of not using PPE can be immediate or long-term. Immediate problems are more obvious, as you may injure yourself. The longer-term consequences could be ill health in the future. If your employer has provided PPE, you have a legal responsibility to wear it.

FIRE AND EMERGENCY PROCEDURES

If there is a fire or an emergency, it is vital that you raise the alarm quickly. You should leave the building or site and then head for the **assembly point.**

When there is an emergency a general alarm should sound. If you are working on a larger and more complex construction site, evacuation may begin by evacuating the area closest to the emergency. Areas will then be evacuated one-by-one to avoid congestion of the escape routes.

Three elements essential to creating a fire

Three ingredients are needed to make something combust (burn):

* oxygen   * heat   * fuel.

The fuel can be anything which burns, such as wood, paper or flammable liquids or gases, and oxygen is in the air around us, so all that is needed is sufficient heat to start a fire.

The fire triangle represents these three elements visually. By removing one of the three elements the fire can be prevented or extinguished.

Figure 1.15 Assembly point sign

How fire is spread

Fire can easily move from one area to another by finding more fuel. You need to consider this when you are storing or using materials on site, and be aware that untidiness can be a fire risk. For example, if there are wood shavings on the ground the fire can move across them, burning up the shavings.

Figure 1.16 The fire triangle

Heat can also transfer from one source of fuel to another. If a piece of wood is on fire and is against or close to another piece of wood, that too will catch fire and the fire will have spread.

On site, fires are classified according to the type of material that is on fire. This will determine the type of fire-fighting equipment you will need to use. The five different types of fire are shown in Table 1.13.

Class of fire	Fuel or material on fire
A	Wood, paper and textiles
B	Petrol, oil and other flammable liquids
C	LPG, propane and other flammable gases
D	Metals and metal powder
E	Electrical equipment

Table 1.13 Different classes of fire

There is also F, cooking oil, but this is less likely to be found on site, except in a kitchen.

Taking action if you discover a fire and fire evacuation procedures

During induction, you will have been shown what to do in the event of a fire and told about assembly points. These are marked by signs and somewhere on the site there will be a map showing their location.

If you discover a fire you should:

* sound the alarm

* not attempt to fight the fire unless you have had fire marshal training

* otherwise stop work, do not collect your belongings, do not run, and do not re-enter the site until the all clear has been given.

Different types of fire extinguishers

Extinguishers can be effective when tackling small localised fires. However, you must use the correct type of extinguisher. For example, putting water on an oil fire could make it explode. For this reason, you should not attempt to use a fire extinguisher unless you have had proper training.

When using an extinguisher it is important to remember the following safety points:

* Only use an extinguisher at the early stages of a fire, when it is small.

* The instructions for use appear on the extinguisher.

* If you do choose to fight the fire because it is small enough, and you are sure you know what is burning, position yourself between the fire and the exit, so that if it doesn't work you can still get out.

Type of fire risk	Fire class Symbol	White label Water	Cream label Foam	Black label Carbon dioxide	Blue label Dry powder	Yellow label Wet chemical
A – Solid (e.g. wood or paper)	A	✓	✓	✗	✓	✓
B – Liquid (e.g. petrol)	B	✗	✓	✓	✓	✗
C – Gas (e.g. propane)	C	✗	✗	✓	✓	✗
D – Metal (e.g. aluminium)	D METAL	✗	✗	✗	✓	✗
E – Electrical (i.e. any electrical equipment)	E	✗	✗	✓	✓	✗
F – Cooking oil (e.g. a chip pan)	F	✗	✗	✗	✗	✓

Table 1.14 Types of fire extinguishers

There are some differences you should be aware of when using different types of extinguisher:

* CO_2 extinguishers – do not touch the nozzle; simply operate by holding the handle. This is because the nozzle gets extremely cold when ejecting the CO_2, as does the canister. Fires put out with a CO_2 extinguisher may reignite, and you will need to ventilate the room after use.

* Powder extinguishers – these can be used on lots of kinds of fire, but can seriously reduce visibility by throwing powder into the air as well as on the fire.

SIGNS AND SAFETY NOTICES

In a well-organised working environment safety signs will warn you of potential dangers and tell you what to do to stay safe. They are used to warn you of hazards. Their purpose is to prevent accidents. Some will tell you what to do (or not to do) in particular parts of the site and some will show you where things are, such as the location of a first aid box or a fire exit.

Types of signs and safety notices

There are five basic types of safety sign, as well as signs that are a combination of two or more of these types. These are shown in Table 1.15.

Type of safety sign	What it tells you	What it looks like	Example
Prohibition sign	Tells you what you must *not* do	Usually round, in red and white	Do not use ladder
Hazard sign	Warns you about hazards	Triangular, in yellow and black	Caution Slippery floor
Mandatory sign	Tells you what you *must* do	Round, usually blue and white	Masks must be worn in this area
Safe condition or information sign	Gives important information, e.g. about where to find fire exits, assembly points or first aid kit, or about safe working practices	Green and white	First aid
Firefighting sign	Gives information about extinguishers, hydrants, hoses and fire alarm call points, etc.	Red with white lettering	Fire alarm call point
Combination sign	These have two or more of the elements of the other types of sign, e.g. hazard, prohibition and mandatory		DANGER Isolate before removing cover

Table 1.15 Different types of safety signs

TEST YOURSELF

1. Which of the following requires you to tell the HSE about any injuries or diseases?

   a. HASAWA

   b. COSHH

   c. RIDDOR

   d. PUWER

2. What is a prohibition notice?

   a. An instruction from the HSE to stop all work until a problem is dealt with

   b. A manufacturer's announcement to stop all work using faulty equipment

   c. A site contractor's decision not to use particular materials

   d. A local authority banning the use of a particular type of brick

3. Which of the following is considered a major injury?

   a. Bruising on the knee

   b. Cut

   c. Concussion

   d. Exposure to fumes

4. If there is an accident on a site who is likely to be the first to respond?

   a. First aider

   b. Police

   c. Paramedics

   d. HSE

5. Which of the following is a summary of risk assessments and is used for high risk activities?

   a. Site notice board

   b. Hazard book

   c. Monitoring statement

   d. Method statement

6. Some substances are combustible. Which of the following are examples of combustible materials?

   a. Adhesives

   b. Paints

   c. Cleaning agents

   d. All of these

7. What is dermatitis?

   a. Inflammation of the skin

   b. Inflammation of the ear

   c. Inflammation of the eye

   d. Inflammation of the nose

8. Screens, netting and guards on a site are all examples of which of the following?

   a. PPE

   b. Signs

   c. Perimeter safety

   d. Electrical equipment

9. Which of the following are also known as scissor lifts or cherry pickers?

   a. Bench saws

   b. Hand-held power tools

   c. Cement additives

   d. Mobile elevated work platforms

10. In older properties the neutral electricity wire is which colour?

    a. Black

    b. Red

    c. Blue

    d. Brown

Chapter 2

Unit CSA–L2Core04

UNDERSTAND INFORMATION, QUANTITIES AND COMMUNICATION WITH OTHERS

LEARNING OUTCOMES

LO1: Know how to interpret and produce information relating to construction

LO2: Understand how to estimate quantities of resources

LO3: Understand how to communicate workplace requirements efficiently

INTRODUCTION

The aim of this chapter is to:

* help you interpret and produce information relating to construction

* show you how to estimate quantities of resources

* enable you to communicate workplace requirements effectively to all levels of the construction team.

INTERPRETING AND PRODUCING INFORMATION

Even quite simple construction projects will require documents. These provide you with the necessary information that you will need to do the job. The documents are produced by a range of different people and each document has a different purpose. Together they give you the full picture of the job, from the basic outline through to the technical specifications.

Types of supporting information

Supporting information can be found in a variety of different types of documents. These include:

* drawings and plans

* programmes of work

* procedures

* specifications

* policies

* schedules

* manufacturers' technical information

* organisational documentation

* training and development records

* risk and method statements

* Construction (Design and Management) (CDM) Regulations

* Building Regulations.

Drawings and plans

Drawings are an important part of construction work. You will need to understand how drawings provide you with the information required to carry out the work. The drawings show what the building will look like and how it will be constructed. This means that there are several different drawings of the building from different viewpoints. In practice, most of the drawings are shown on the same sheet.

Block plans

Block plans show the construction site and the surrounding area. Normally block plans are at a ratio of 1:2500 and 1:1250. This means that 1 millimetre on a block plan is equal to 2,500 mm (2.5 m) or 1,250 mm (1.25 m) on the ground.

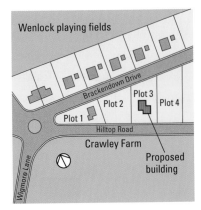

Figure 2.1 Block plan

Site plan

Location drawings are sometimes known as site plans. The site plan drawing shows what is basically planned for the site. It is an important drawing because it has been created in order to get approval for the project from planning committees or funding sources. In most cases the site plan is actually an architectural plan, showing the basic arrangement of buildings and any landscaping.

The site plan will usually show:

* directional orientation (i.e. the north point)

* location and size of the building or buildings

* existing structures

* clear measurements

* colours and materials to be used.

General location

Location drawings show the site or building in relation to its surroundings. It will therefore show details such as boundaries, other buildings and roads. It will also contain other vital information, including:

* access

* drainage

* sewers

* the north point.

The drawing will have a title and will show the scale. A job or project number will help to identify it easily, and it will also have an address, the date when the drawing was done and the name of the client. A version number will also be on the drawing, with an amendment date if there have been any changes. It is important to make sure you have the latest drawing.

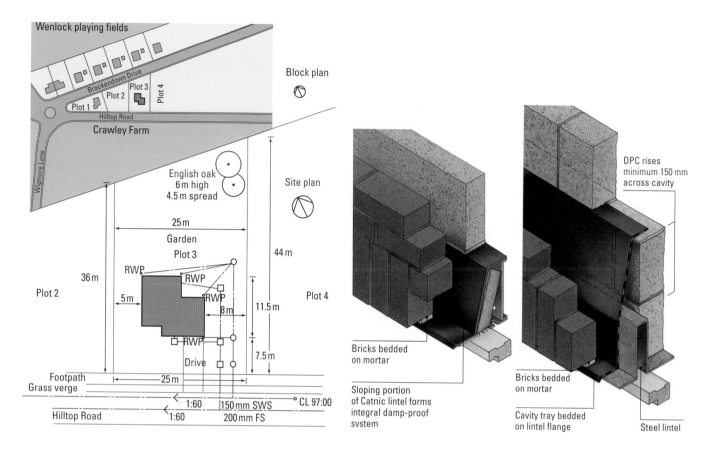

Figure 2.2 Location plan

Figure 2.3 Assembly drawing

Normally location drawings are either 1:500 or 1:200 (that is, 1 mm of the drawing represents 500 mm or 200 mm on the ground).

Assembly

These are detailed drawings that illustrate the different elements and components of the construction. They are likely to be 1:20, 1:10 or 1:5 (1 mm of the drawing represents 20 mm, 10 mm or 5 mm on the ground). This larger scale allows more detail to be shown, to ensure accurate construction.

Sectional

These drawings aim to provide:

* vertical dimensions

* constructional details

* horizontal sections.

They can be used to show the height of ground levels, damp-proof courses, foundations and other aspects of the construction.

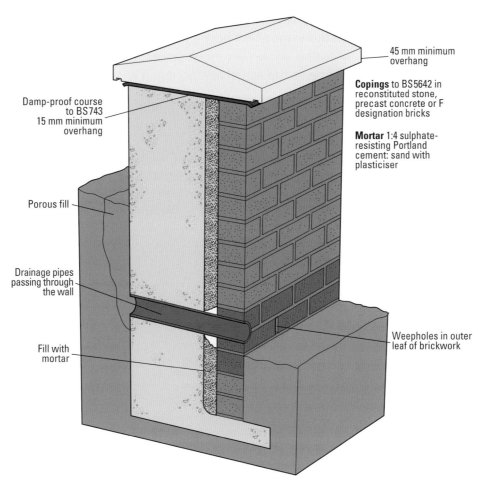

45 mm minimum overhang

Copings to BS5642 in reconstituted stone, precast concrete or F designation bricks

Damp-proof course to BS743 15 mm minimum overhang

Mortar 1:4 sulphate-resisting Portland cement: sand with plasticiser

Porous fill

Drainage pipes passing through the wall

Weepholes in outer leaf of brickwork

Fill with mortar

Figure 2.4 Section drawing of an earth retaining wall

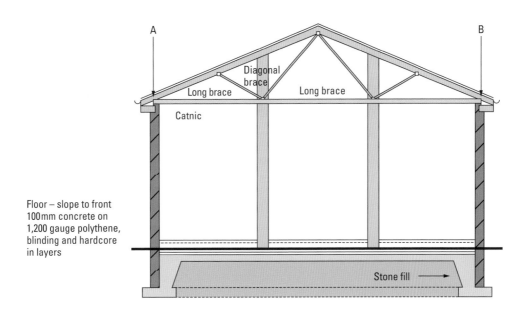

A

B

Diagonal brace

Long brace

Long brace

Catnic

Floor – slope to front 100mm concrete on 1,200 gauge polythene, blinding and hardcore in layers

Stone fill →

Figure 2.5 Section drawing of a garage

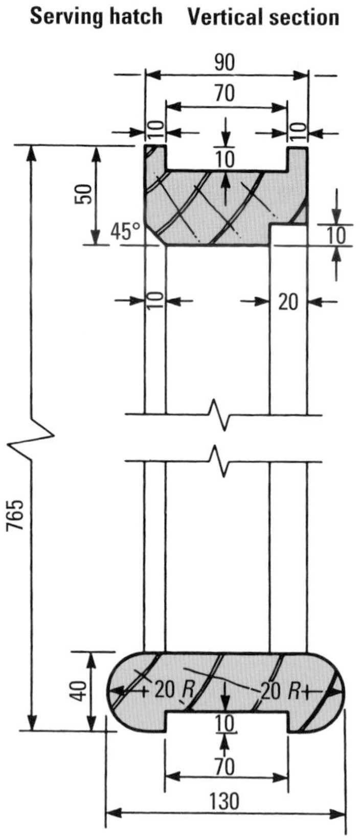

Serving hatch   Vertical section

Figure 2.6 Detail drawing

Details

These drawings show how a component needs to be manufactured. They can be shown in various scales, but mainly 1:10, 1:5 and 1:1 (the same size as the actual component if it is small).

Orthographic projection (first angle)

First angle projection is a view that represents the side of the object as if you were standing away from it, as can be seen in Fig 2.7.

Isometric projection

Isometric projection is a way of representing three-dimensional objects in two dimensions, as can also be seen in Fig 2.7. All horizontal lines are drawn at 30°.

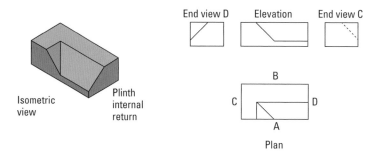

Figure 2.7 First angle projection

Programmes of work

Programmes of work show the actual sequence of any work activities on a construction project. Part of the work programme plan is to show target times. They are usually shown in the form of a bar or Gantt chart (a special kind of bar chart), as can be seen in Fig 2.8.

In this figure:

* on the left hand side all of the tasks are listed – note this is in logical order

* on the right the blocks show the target start and end date for each of the individual tasks

* the timescale can be days, weeks or months.

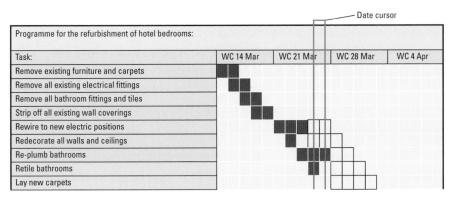

Figure 2.8 Single line contract plan Gantt chart

Far more complex forms of work programmes can also be created. The Gantt chart shown below (Fig 2.9) shows the construction of a house.

This is a more complex example of a bar chart:

* There are two lines – they show the target dates and actual dates. The actual dates are shaded, showing when the work actually began and how long it actually took.

* If this bar chart is kept up to date an accurate picture of progress and estimated completion time can be seen.

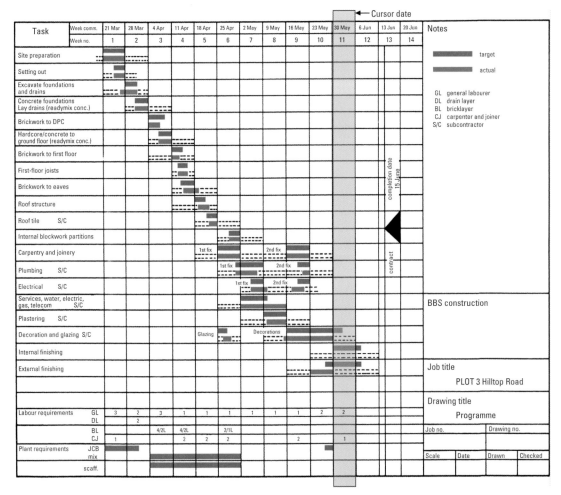

Figure 2.9 Gantt chart for the construction of a house

Procedures

When you work for a construction company it will have a series of procedures which you will have to follow. A good example is the emergency procedure. This will explain precisely what is required in the case of an emergency on site and who will have responsibility for carrying out particular duties. Procedures are there to show you the right way of doing something.

Another good example of a procedure is the procurement or buying procedure. This will outline:

* who is authorised to buy what, and how much individuals are allowed to spend

* any forms or documents that have to be completed when buying.

Specifications

In addition to drawings it is usually necessary to have documents known as specifications. These provide much more information, as can be seen in Fig 2.10.

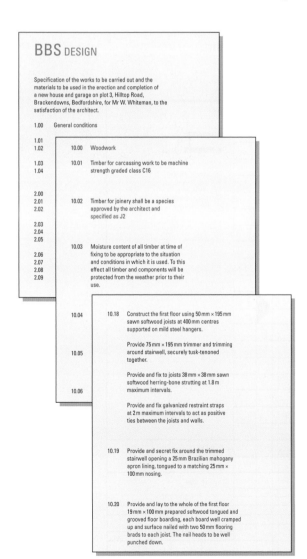

BBS DESIGN

Specification of the works to be carried out and the materials to be used in the erection and completion of a new house and garage on plot 3, Hilltop Road, Brackendowns, Bedfordshire, for Mr W. Whiteman, to the satisfaction of the architect.

1.00   General conditions

1.01
1.02   10.00   Woodwork

1.03   10.01   Timber for carcassing work to be machine
1.04   strength graded class C16

2.00
2.01   10.02   Timber for joinery shall be a species
2.02   approved by the architect and
                     specified as J2
2.03
2.04
2.05
            10.03   Moisture content of all timber at time of
2.06   fixing to be appropriate to the situation
2.07   and conditions in which it is used. To this
2.08   effect all timber and components will be
2.09   protected from the weather prior to their
                     use.

    10.04   10.18   Construct the first floor using 50 mm × 195 mm
                      sawn softwood joists at 400 mm centres
                      supported on mild steel hangers.

                      Provide 75 mm × 195 mm trimmer and trimming
    10.05   around stairwell, securely tusk-tenoned
                      together.

                      Provide and fix to joists 38 mm × 38 mm sawn
                      softwood herring-bone strutting at 1.8 m
    10.06   maximum intervals.

                      Provide and fix galvanized restraint straps
                      at 2 m maximum intervals to act as positive
                      ties between the joists and walls.

    10.19   Provide and secret fix around the trimmed
             stairwell opening a 25 mm Brazilian mahogany
             apron lining, tongued to a matching 25 mm ×
             100 mm nosing.

    10.20   Provide and lay to the whole of the first floor
             19 mm × 100 mm prepared softwood tongued and
             grooved floor boarding, each board well cramped
             up and surface nailed with two 50 mm flooring
             brads to each joist. The nail heads to be well
             punched down.

Figure 2.10 Extracts from a typical specification

The specifications give you a precise description. They will include:

* the address and description of the site

* on-site services (e.g. water and electricity)

* materials description, outlining the size, finish, quality and tolerances

* specific requirements, such as the individual who will authorise or approve work carried out

* any restrictions on site, such as working hours.

Policies

Policies are sets of principles or a programme of actions. These are two good examples:

* Environmental policy – how the business goes about protecting the environment.

* Safety policy – how the business deals with health and safety matters and who is responsible for monitoring and maintaining it.

You will normally find both policies and procedures in site rules. These are usually explained to each new employee when they first join the company. Sometimes there may be additional site rules, depending on the job and the location of the work.

Schedules

Schedules are cross-referenced to drawings that have been prepared by an architect. They will show specific design information. Usually they are prepared for jobs that will be carried out regularly on site, such as:

⁕ working on windows, doors, floors, walls or ceilings

⁕ working on drainage, lintels or sanitary ware.

A schedule can be seen in Fig 2.11.

The schedule is very useful for a number of purposes, such as:

⁕ working out the quantities of materials needed

⁕ ordering materials and components and then checking them against deliveries

⁕ locating where specific materials will be used.

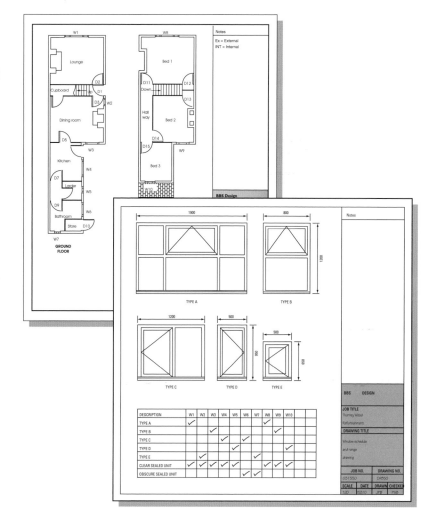

Figure 2.11 Typical windows schedule, range drawing and floor plans

Manufacturers' technical information

Almost everything that is bought to be used on site will come with a variety of types of information. The basic technical information provided will show what the equipment or material is intended to be used for, how it should be stored and any particular requirements it may have, such as for handling or maintenance.

Technical information from the manufacturer can come from a variety of different sources. These may include:

⁕ printed or downloadable data sheets

⁕ printed or downloadable user instructions

⁕ manufacturers' catalogues or brochures

⁕ manufacturers' websites.

Organisational documentation

The potential list of organisational documentation and paperwork is extensive. These are outlined in Table 2.1. Examples can be seen in Figs 2.12 to 2.16.

Document	Purpose
Timesheet	Record of hours that you have worked and the jobs that you have carried out. This is used to help work out your wages and the total cost of the job.
Day worksheet	This details work that has been carried out without providing an estimate beforehand. It usually includes repairs or extra work and alterations.
Variation order	Provided by the architect and given to the builder, showing any alterations, additions or omissions to the original job.
Confirmation notice	Provided by the architect to confirm any verbal instructions.
Daily report or site diary	This covers things that might affect the project like detailed weather conditions, late deliveries or site visitors.
Orders and requisitions	These are order forms, requesting the delivery of materials.
Delivery notes	These are provided by the supplier of materials as a list of all materials being delivered. These need to be checked against materials actually delivered. The buyer will sign the delivery note when they are happy with the delivery.
Delivery records	These are lists of all materials that have been delivered on site.
Memorandum	These are used for internal communications and are usually brief.
Letters	These are used for external communications, usually to customers or suppliers.
Fax	Even though email is commonly used, the industry still uses faxes, as they provide an exact copy of an original document.

Table 2.1 Types of organisational documentation

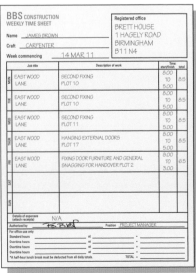

Figure 2.12 Timesheet

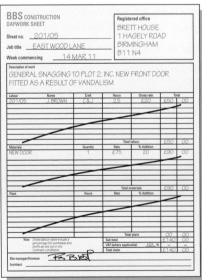

Figure 2.13 Day worksheet

Figure 2.14 Variation order

Training and development records

Training and development is an important part of any job, as it ensures that employees have all the skills and knowledge that they need to do their work. Most medium to large employers will have training policies that set out how they intend to do this.

Employers will have a range of different documents to keep records and to make sure that they are on track. These documents will record all the training that an employee has undertaken.

Training can take place in a number of different ways and different places. It can include:

* induction

* toolbox talks

* in-house training

* specialist training

* training or education leading to formal qualifications.

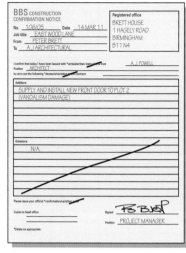

Figure 2.15 Confirmation notice

Checking information for conformity

The information to be checked can include drawings, programmes of work, schedules, policies, procedures, specifications and so on. The term 'conformity' in this sense means:

* making sure that any part of the assembly or component is suitable for the job

* making sure that the standard of work meets the necessary performance requirements.

This may mean that there could be an industry or trade standard that will need to be followed. The actual job or client may also require specific standards.

Figure 2.16 Daily report or site diary

Interpreting construction specifications

It would be difficult to put in all of the details in full, so symbols, hatchings and abbreviations are used to simplify the drawings. All of these symbols or hatchings are drawn to follow BS1192. The symbols cover various types of brickwork and blockwork, as well as concrete, hard core and insulation, as can be seen in Fig 2.17.

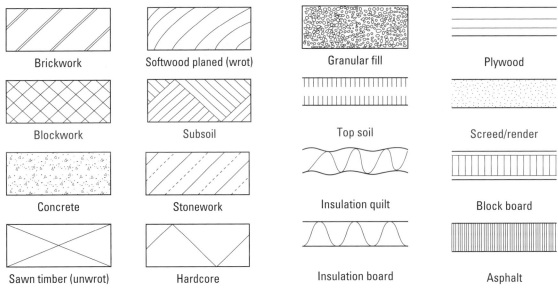

Figure 2.17 Symbols used on drawings

Abbreviation	Meaning
bwk	Refers to all types of brickwork
conc	Refers to areas that will be concreted
dpc	Refers to all types of damp-proof course
fdn	Refers to foundations that are required
insul	Refers to location and type of insulation
rwg	Refers to location of rainwater gulleys
svp	Refers to location and type of soil and vent pipe

Table 2.2 Abbreviations used in construction drawings

Common abbreviations

As we have already seen, covering the drawing with full detail would make it hard to read, so abbreviations are used. Table 2.2 outlines some examples of abbreviations that you will need to become familiar with.

Drawing equipment and its uses

Some basic equipment is necessary in order to produce drawings. These items are outlined in Table 2.3.

Equipment	Explanation and use
Scale rule	This is an essential piece of equipment. It needs to have 1:5/1:50, 1:10/1:100, 1:20/1:200 and 1:250/1:2500.
Set square	You will need to have a pair of these, or an adjustable square. If it is adjustable then you need to be able to create angles of up to 90°. The set square on the shortest side should be at least 150 mm. You will need the ability to create 30, 45, 60 and 90° angles.
Protractor	A protractor is essential to be able to measure angles up to and including 180°.
Compass	Compasses are used to create circles or arcs. It is also advisable to have a divider so that you can easily transfer measurements and dividing lines.
Pencils	For drawings you will need a 2H, 3H or 4H pencil. For sketching and darkening outlines you will need an HB pencil. You will need to keep these sharp.

Table 2.3 Drawing equipment required

In addition to this you will also need at least an A2 size drawing board that has a parallel rule (these may be provided by your college). It is also useful to have an eraser.

Scales used to produce construction drawings

When the plans for individual buildings or construction sites are drawn up they have to be scaled down so that they will fit on a manageable size of paper. It is important to remember that drawings are not sketches and that they are drawn to scale. This means that they are:

* exact and accurate

* in proportion to the real construction.

You can work out the dimensions by using the scale rule when measuring the drawings. There are several common scales used and the measurement is usually metric:

* 1:2500 – the drawing is 2,500 times smaller than the real object

* 1:100 – the drawing is 100 times smaller than the real object

* 1:50 – the drawing is 50 times smaller than the real object

* 1:20 – the drawing is 20 times smaller than the real object

* 1:10 – the drawing is 10 times smaller than the real object

* 1:5 – the drawing is 5 times smaller than the real object

* 1:2 – the drawing is 2 times smaller than the real object.

ESTIMATING QUANTITIES OF RESOURCES

Working out the quantity and cost of resources that are needed to do a particular job is, perhaps, one of the most difficult tasks. In most cases you or the company you work for will be asked to provide a price for the work.

It is generally accepted that there are three ways of doing this:

* estimate – an approximate calculation based on available information

* quotation – which is a fixed price

* tender – tendering is a process of allowing various parties to price for the same work. The process can be open or closed. This usually means that the result is fair.

As we will see a little later in this section, these three ways of costing are very different and each of them has its own problems.

Methods used to estimate quantities

Obviously past experience will help you to quickly estimate the amount of materials that will be needed on particular construction projects. This is also true of working out the best place to buy materials and how much the labour costs will be to get the job finished.

Many businesses will use the *Hutchins UK Building Costs Blackbook*, which provides a construction cost guide. It breaks down all types of work and shows an average cost for each of them.

Computerised estimating packages are available, which will give a comprehensive detailed estimate that looks very professional. This will also help to estimate quantities and timescales.

The alternative is of course to carry out a numerical calculation. It is therefore important to have the right resources upon which to base these calculations. These could be working drawings, schedules or other documents.

Long lengths in kilometres (km)

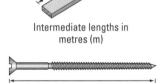

Intermediate lengths in metres (m)

Usually all this involves making additions, subtractions, multiplications and divisions. In order to work out the amount of materials you will need for a construction project you will need to know some basic information:

* What does the job entail? How complex is it, and how much labour is required?

* What materials will be used?

* What are the costs of the materials?

Measurement

The standard unit for measurement is the metre (m). There are 100 centimetres (cm) and 1,000 millimetres (mm) in a metre. It is important to remember that drawings and plans have different scales, so these need to be converted to work out the quantities of materials required.

Small lengths in millimetres (mm)

Figure 2.18 Length in metres and millimetres

The most basic thing to work out is length (see Fig 2.18), from which you can calculate perimeter, area and then volume, capacity, mass and weight, as can be seen in Table 2.4.

Measurement	Explanation
Length	This is the distance from one end to the other. This could be measured in metres or millimetres, depending on the job.
Perimeter	This helps you work out the distance around a shape, such as the size of a room or a garden. It will help you estimate the length of a wall, for example. You just need to measure each side and then add them together (see Fig 2.20).
Area	You can work out the area of a room, for example, by measuring the length and the width of the room. Then you multiply the width by the length to give the number of square metres (m^2) (see Fig 2.20).
Volume and capacity	Volume shows how much space is taken up by an object, such as a room. Again this is simply worked out by multiplying the width of the room by its length and then by its height. This gives you the number of cubic metres (m^3). Capacity works in exactly the same way but instead of showing the figure as cubic metres you show it as litres. This is ideal if you are trying to work out the capacity of a water tank or a garden pond (see Fig 2.19).
Mass or weight	Mass is measured usually in kilograms or in grammes. Mass is the actual weight of a particular object, such as a brick.

Table 2.4 Working out measurements

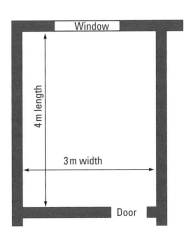

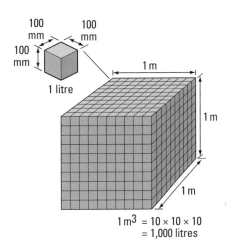

$1 m^3 = 10 \times 10 \times 10$
$= 1,000$ litres

Figure 2.19 Measuring area and perimeter

Figure 2.20 Relationship between volume and capacity

Formulae

These can appear to be complicated, but using formulae is essential for working out quantities of materials. Each of the formulae is related to different shapes. In construction work you will often have to work out quantities of materials needed for odd shaped areas.

Area

To work out the area of a triangular shape, you use the following formula:

$$\text{Area (A)} = \text{Base (B)} \times \frac{\text{Height (H)}}{2}$$

So if a triangle has a base of 4.5 and a height of 3.5 the calculation is:

$$4.5 \times \frac{3.5}{2}$$

$$\text{Or } 4.5 \times 3.5 = \frac{15.75}{2} = 7.875 \, m^2$$

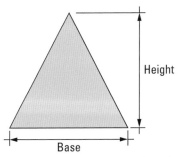

Figure 2.21 Triangle

Height

If you want to work out the height of a triangle you switch the formulae around. To give:

$$\text{Height} = 2 \times \frac{\text{Area}}{\text{Base}}$$

Perimeter

To work out the perimeter of a rectangle you use the formula:

$$\text{Perimeter} = 2 \times (\text{Length} + \text{Width})$$

It is important to remember this because you need to count the length and the width twice to ensure you have calculated the total distance around the object.

Circles

To work out the circumference or perimeter of a circle you use the formula:

$$\text{Circumference} = \pi \, (\text{pi}) \times \text{Diameter}$$

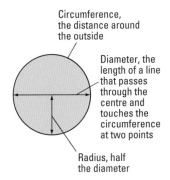

Figure 2.22 Parts of a circle

π (pi) is always the same for all circles and is 3.142.

Diameter is the length of the widest part.

If you know the circumference and need to work out the diameter of the circle the formula is:

$$\text{Diameter} = \frac{\text{Circumference}}{\pi \text{ (pi)}}$$

For example if a circle has a circumference of 15.39 m then to work out the diameter:

$$\frac{15.39}{3.142} = 4.89\,\text{m}$$

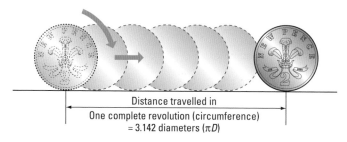

Distance travelled in
One complete revolution (circumference)
= 3.142 diameters (πD)

Figure 2.23 Relationship between circumference and diameter

Complex areas

Land, for example, is rarely square or rectangular. It is made up of odd shapes. You should never feel overwhelmed by complex areas, as all you need to do is to break them down into regular shapes.

By accurately measuring the perimeter you can then break down the shape into a series of triangles or rectangles. All that you need to do then is to work out the area of each of the shapes within the overall shape and add them up together.

Shape		Area equals	Perimeter equals
Square		AA (or A multiplied by A)	4A (or A multiplied by 4)
Rectangle		LB (or L multiplied by B)	2(L+B) (or L plus B multiplied by 2)

Shape	Area equals	Perimeter equals
Trapezium	$\dfrac{(A + B)H}{2}$ (or A plus B multiplied by H and then divided by 2)	A + B + C + D
Triangle	$\dfrac{BH}{2}$ (or B multiplied by H and then divided by 2)	A + B + C
Circle	πr^2 (or r multiplied by itself and then multiplied by pi (3.142))	πd or $2\pi r$

Table 2.5 Calculating complex areas

Volume

Sometimes it is necessary to work out the volume of an object, such as a cylinder or the amount of concrete needed. All that needs to be done is to work out the base area and then multiply that by the height.

For a concrete volume, if a 1.2 m square needs 3 m of height then the calculation is:

$$1.2 \times 1.2 \times 3 = 4.32 \, \text{m}^3$$

To work out the volume of a cylinder you need to know the base area × the height. The formula is:

$$\pi r^2 \times H$$

So if a cylinder has a radius (r) of 0.8 and a height of 3.5 m then the calculation is:

$$3.142 \times 0.8 \times 0.8 \times 3.5 = 7.038 \, \text{m}^3$$

Pythagoras

Pythagoras' theorem is used to work out the length of the sides of right-angled triangles. It states that:

In all right-angled triangles the square of the longest side is equal to the sum of the squares of the other two sides (that is, the length of a side multiplied by itself).

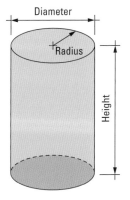

Figure 2.24 Cylinder

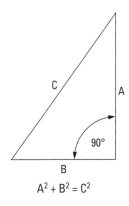

$$A^2 + B^2 = C^2$$

Figure 2.25 Pythagoras' theorem

Measuring materials

Using simple measurements and formulae can help you work out the amount of materials you will need. This is all summarised in Table 2.6.

Material	Measurement
Timber	To work out the linear run of a cubic metre of timber of a given cross sectional area, divide a square metre by the cross sectional area of one piece.
Flooring	To work out the amount of flooring for a particular area in metres^2 multiply the width of the floor by the length of the floor.
Stud walling, rafters and joists	Measure the distance that the stud partition will cover then divide that distance by a specified spacing and add 1. This will give you the number of spaces between each stud.
Fascias, barges and soffits	Measure the length and then add 10% for waste; however, this will depend on the nearest standard metric size of timber available.
Skirting, dado, picture rails and coving	You need to work out the perimeter of the room and then subtract any doorways or other openings. Again, add 10% for waste.
Bricks and mortar	Half brick walls use 60 bricks per metre squared and one brick walls use double that amount. You should add 5 per cent to take into account any cutting or damage. For mortar assume that you will need 1 kg for each brick.

Table 2.6 Working out materials required

How to cost materials

Once you have found out the quantity of materials necessary, you need to find out the price of those materials. You then do the costing by simply multiplying those prices by the amount of materials actually needed.

CASE STUDY

South Tyneside Homes

South Tyneside Council's Housing Company

It's important to get it right

Glen Campbell is a team leader at South Tyneside Homes.

'Your English and maths skills really are important. As an apprentice, you have to be able to communicate properly – to get information and materials back and forth between tradespeople and yourself, to be able to sit and put a little drawing down, to label things up, and to take information off drawings – especially on the capital works jobs. You're reading and writing stuff down all the time… even your timesheets because they have to be accurate.

When it comes to your maths skills, you're using measurement all the time. If you get measurements wrong, you're not making the money. For example, if you're using the wrong size timber for a roof – the drawing says you've got to use 200 × 50 mm joists and then you go and use ones that are 150 mm – it's either going to cost you more to go back and get it right, or it's not going to be able to take that stress load once the roof goes on. In the end it could even collapse.'

Materials and purchasing systems

Many builders and companies will have preferred suppliers of materials. Many of them will already have negotiated discounts based on their likely spending with that supplier over the course of a year. The supplier will then be organised to supply them at an agreed price.

In other cases, builders may shop around to find the best price for the materials that match the specification. The lowest price may not necessarily be the best one to go for. All materials need to be of a sufficient quality. The other key consideration is whether the materials are immediately available for delivery.

It is vital that suppliers are reliable and that they have sufficient materials in stock. Delays in deliveries can cause major setbacks on site. It is not always possible to warn suppliers that materials will be needed, but a well-run site should be able to anticipate the materials that are needed and put in the orders within good time.

Large quantities may be delivered direct from the manufacturer straight to site. This is preferable when dealing with items where colours must be consistent.

Comparing estimated labour rates

The cost of labour for particular jobs is based on the hourly charge-out rate for that individual or group of individuals multiplied by the time it would take to complete the job.

Labour rates can depend on:

* the expertise of the construction worker

* the size of the business they work for

* the part of the country in which the work is being carried out

* the complexity of the work.

According to the International Construction Costs Survey 2012, the following were average costs per hour:

* Group 1 tradespeople – plumbers, electricians etc. – £30

* Group 2 tradespeople – carpenters, bricklayers etc. – £30

* Group 3 tradespeople – tillers, carpet layers and plasterers – £30

* General labourers – £18

* Site supervisors – £46.

Quotes, estimated prices and tenders

As we have already seen, estimates, quotes and tenders are very different. It is useful to look at these in slightly more detail, as can be seen in Table 2.7 below.

Type of costing	Explanation
Estimate	This needs to be a realistic proposal of how much a job will cost. An estimate is not binding and the client needs to understand that the final cost might be more.
Quote	This is a fixed price based on a fixed specification. The final price may be different if the fixed specification changes, for example if the customer asks for additional work then the price will be higher.
Tender	This is a competitive process. The customer advertises the fact that they want a job done and invites tenders. The customer will specify the specifications and schedules and may even provide the drawings. The companies tendering then prepare their own documents and submit their price based on the information the customer has given them. All tenders are submitted to the customer by a particular date and are either open or closed. The customer then opens all tenders on a given date and awards the contract to the company of their choice. This process is particularly common among public sector customers, such as local authorities.

Table 2.7 Estimates, quotes and tenders

Implications of inaccurate estimates

Larger companies will have an estimating team. Smaller businesses will have someone who has the job of being an estimator. Whenever they are pricing a job, whether it is a quote, an estimate or a tender, they will have to work out the costs of all materials, labour and other costs. They will also have to include a **mark-up**.

It is vital that all estimating is accurate. Everything needs to be measured and checked. All calculations need to be double-checked.

It can be disastrous if these figures are wrong because:

* if the figure is too high then the client is likely to reject the estimate and look elsewhere as some competitors could be cheaper

* if the figure is too low then the job may not provide the business with sufficient profit and it will be a struggle to make any money out of the job.

COMMUNICATING WORKPLACE REQUIREMENTS

Communication can be split into two different types:

* Verbal communication – including face-to-face conversations, discussions in meetings or performance reviews and talking on the telephone.

* Written communication – including all forms of documents, from letters and emails to drawings and work schedules.

Each of these forms of communication needs to be clear, accurate and designed in such a way as to make sure that whoever has to use it or refer to it understands it.

Figure 2.26 It's important to communicate effectively, whether it's verbal or written

Key personnel in the communication cycle

Each construction job will require the services of a team of professionals. They have to be able to work and communicate effectively with one another. Each team has different roles and responsibilities. They can be broken down into three particular groups:

* on site   * off site   * visitors.

These are described in Tables 2.8, 2.9 and 2.10.

Role	Responsibilities
Apprentices	They can work for any of the main building services trades under supervision. They only carry out work that has been specifically assigned to them by a trainer, a skilled operative or a supervisor.
Skilled or trade operative	A specialist in a particular trade, such as bricklaying or carpentry. They will be qualified in that trade, or working towards their qualification
Unskilled operatives	Also known as labourers, these are entry level operatives without any formal training. They may be experienced on sites and will take instructions from the supervisor or site manager.
Building services engineers	They are involved in the design, installation and maintenance of heating, water, electrics, lighting, gas and communications. They work either for the main contractor or the architect and give instruction to building services operatives.
Building services operatives	They include all the main trades involved in installation, maintenance and servicing. They take instruction from the building services engineers and work with other individuals, such as the supervisor and charge-hand.
Charge-hand	This person supervises a specific trade, such as carpenters and bricklayers.
Trade foreperson	This person supervises the day-to-day running of the site, and organises the charge-hand and any other operatives.
Site manager	This person runs the construction site, makes plans to avoid problems and meet deadlines, and ensures all processes are carried out safely. They communicate directly with the client.
Supervisor	The supervisor works directly for the site manager on larger projects and carries out some of the site manager's duties on their behalf.
Health and safety officer	This person is responsible for managing the safety and welfare of the construction site. They will carry out inspections, provide training and correct hazards.

Table 2.8 On-site construction team

Role	Responsibilities
Client	The client, such as a local authority, commissions the job. They define the scope of the work and agree on the timescale and schedule of payments.
Customer	For domestic dwellings, the customer may be the same as the client, but for larger projects a customer may be the end user of the building, such as a tenant renting local authority housing or a business renting an office. These individuals are most affected by any work on site. They should be considered and informed with a view to them suffering as little disruption as possible.
Architect	They are involved in designing new buildings, extensions and alterations. They work closely with clients and customers to ensure the designs match their needs. They also work closely with other construction professionals, such as surveyors and engineers.
Consultant	Consultants such as civil engineers work with clients to plan, manage, design or supervise construction projects. There are many different types of consultant, all with particular specialisms.
Main contractor	This is the main business or organisation employed to head up the construction work. The contractor organises the on-site building team and pulls together all necessary expertise. They manage the whole project, taking full responsibility for its progress and costs.
Clerk of works	This person is employed by the architect on behalf of a client. They oversee the construction work and ensure that it represents the interests of the client and follows agreed specifications and designs.
Quantity surveyor	Quantity surveyors are concerned with building costs. They balance maintaining standards and quality against minimising the costs of any project. They need to make choices in line with Building Regulations. They may work either for the client or for the contractor.
Estimator	Estimators calculate detailed cost breakdowns of work based on specifications provided by the architect and main contractor. They work out the quantity and costs of all building materials, plant required and labour costs.
Sub-contractor	They carry out work on behalf of the main contractor and are usually specialist tradespeople or professionals, such as electricians. Essentially, they provide a service and are contracted to complete their part of the project.
Supplier/wholesaler contracts manager	They work for materials suppliers or stockists, providing materials that match required specifications. They agree prices and delivery dates.

Table 2.9 Off-site construction team

Site visitor	Role and responsibility
Training officers and assessors	These people work for approved training providers. They visit the site to observe and talk to apprentices and their mentors or supervisors. They assess apprentices' competence and help them to put together the paperwork needed to show evidence of their skills.
Building control inspector	This person works for the local authority to ensure that the construction work conforms to regulations, particularly the Building Regulations. They check plans, carry out inspections, issue completion certificates, work with architects and engineers and provide technical knowledge on site.
Water inspector	This person carries out checks of plumbing and drainage systems on construction sites.
Health and Safety Executive (HSE) inspector	An HSE inspector can enter any workplace without giving notice. They will look at the workplace, the activities and the management of health and safety to ensure that the site complies with health and safety laws. They can take action if they find there is a risk to health and safety on site.
Electrical services inspector	Inspectors are approved by the National Inspection Council for Electrical Installation Contracting. They check all electrical installation has been carried out in accordance with legislation, particularly Part P of the Building Regulations.

Table 2.10 Construction visitors

Effects of poor communication

Effective communication is essential in all types of work. It needs to be clear and to the point, as well as accurate. Above all it needs to be a two-way process. This means that any communication that you have with anyone must be understood by them. It means thinking before communicating. Never assume that someone understands you unless they have confirmed that they do.

In construction work you have to keep to schedule and work on time, and it is important to follow precise instructions and specifications. Failing to communicate will always cause confusion, extra cost and delays; it can lead to problems with health and safety and accidents. Such problems are unacceptable and very easy to avoid. Negative communication or poor communication can damage the confidence that others have in you to do your job.

Good communication means efficiency and achievement.

> **REED TIP**
>
> As well as within your own team, it is important to communicate clearly with the other trades working on a site, especially if there's a problem that may delay the next stage of the job.

Communication techniques and teamwork

It is important to have a good working relationship with colleagues at work. An important part of this is to communicate in a clear way with them. This helps everyone understand what is going on and what decisions have been made. It also means being clear. Most communication with colleagues will be verbal (spoken). Good communication means:

* cutting out mistakes and stoppages (saving money)

* avoiding delays

* making sure that the job is done right the first time and every time.

Figure 2.27 A water inspection

Equality and diversity in communication

Equality and diversity is not simply about treating everyone in the same way. It is actually recognising that people are different and have different needs. Each of us is unique. This could mean that you are working with people of a different culture, a different age (younger or older), or who follow different religions. It might refer to marital status or gender, sexual orientation or your first language.

In all your actions and your communications you should:

* recognise and respect other people's backgrounds

* recognise that everyone has rights and responsibilities

* not harass or be offensive and use language or behaviour that discriminates.

You should also remember that not everyone's first language will be English so they may not understand everything or be able to communicate clearly with you. You might also find that some colleagues may have hearing impairments (or may not hear what you're saying because they are in a noisy environment). In cases like these, use simple language and check that both you and the person you are communicating with have understood the message.

CASE STUDY

South Tyneside Homes

South Tyneside Council's Housing Company

Using writing and maths in the real world

Gary Kirsop, Head of Property Services, says:

'People seem to think that trades are all about your hands, but it's more than that. You're measuring complicated things – all the trades need to have about the same technical level for planning, calculation and writing reports. You need that level to get through your exams for the future too. When you have one day a week in college, but four days a week working with customers in the real world, without communications skills, it would all fall apart. You have to understand that people come from different backgrounds and that they have their own communication modes. Having good GCSEs will really help you get by in the trade.'

Advantages and disadvantages of different methods of communication

As you progress in your career in construction, you may come across a number of different documents that are used either in the workplace or are provided to customers or clients. All of these documents have a specific purpose. Their exact design may vary from business to business, but the information contained on them will usually be similar.

Documents in the workplace
This group of documents tend to be used only within the workplace. Their general purpose is to collect information or to pass on information from one part of the business to another.

Document type	Purpose
Job specifications	These are detailed sets of requirements that cover the construction, features, materials, finishes and performance specifications required for each major aspect of a project. They may, for example, require a particular level of energy efficiency.
Plans or drawings	These are prepared by architects. They are drawn to scale and provide a standard detailed drawing. They will be used as blueprints (instructions) by building services engineers and operatives while they are working on the site.
Work programmes	These are detailed breakdowns of the order in which work needs to be completed, along with an estimate as to how long each stage is likely to take. For example, a certain amount of time will be allocated for site preparation and then piling and the construction of the substructure of the building. The work programme will indicate when particular skills will be needed and for approximately how long.
Purchase orders	These are documents issued by the buyer to a supplier. They detail the type of materials, quantity and the agreed price so form a record of what has been agreed. The order for materials will have been discussed with the supplier before the purchase order is completed. Many purchase orders are now transmitted electronically, although paper records may be necessary for future reference.
Delivery notes	These are issued to the buyer by the supplier. They act as a checklist for the buyer to ensure that every item requested on the purchase order has been delivered. The buyer will sign the delivery note when they are satisfied with the delivery.
Timesheets	These are completed by those working on site and are verified by the charge-hand, site manager or supervisor. They detail the start and finish times of each individual working on site. They form the basis of the pay calculation for that worker and the overall time that the job has taken.
Policy documents	These cover health and safety, environmental or customer service issues, among others. They outline the requirements of all those working on the site. They will identify roles and responsibilities, codes of conduct or practice, and methods and remedies for dealing with problems or breaches of policy.

Table 2.11 Documents used in the workplace

Documents for customers and clients

Some documents need to be provided to customers and clients. They are necessary to pass on information and can include records of costs and charges that the customer or client is expected to pay for work carried out. Table 2.12 describes what these documents are and their purpose.

Document	Purpose
Quotations and tenders	Quotations provide written details of the costs of carrying out a particular job. They are based on the specification or requirements of the customer or client. They will usually be written by the main contractor on larger sites. A tender is usually a sealed quotation submitted by a contractor at the same time as tenders from other firms in the hope that their quotation will not only match the requirements but will also be the cheapest and therefore the most likely to win the work.
Estimates	An estimate differs from a quotation because it is not a binding quote but a calculation of the cost based on what the contractor thinks the work may involve.
Invoices	An invoice is a list of materials or services that have been provided. Each has an itemised cost and the total is shown at the bottom of the document, along with any additional charges such as VAT.

Document	Purpose
Account statements	This is a record of all the transactions (invoices and payments) made by a customer or client over a given period. It matches payments by the customer and client against invoices raised by the supplier. It also notes any money still owing or over-payments that may have been made.
Contracts	A contract is a legally binding agreement, usually between a contractor and a customer or client, which states the obligations of both parties. A series of agreements are made as part of the contract. It binds both parties to stick to the agreement, which may detail timescales, level of work or costs.
Contract variations	Contract variations are also legally binding. They may be required if both the supplier and the customer or client agrees to change some of the terms of the original contract. This could mean, for example, additional obligations, renegotiating prices or new timescales.
Handover information	Once a project, such as an installation, has been completed, the installer that commissioned the installation will check that it is performing as expected. Handover information includes: • the commissioning document, which details the performance and the checks or inspections that have been made • an installation certificate, which shows that the work has been carried out in accordance with legal requirements and the manufacturer's recommendations.

Table 2.12 Documents used with customers and clients

Other forms of communication
So far we have mainly focused on written forms of communication.

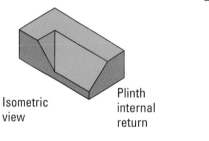
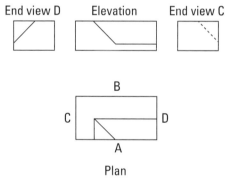

Figure 2.28 An example bill of quantities

However one of the most common forms of communication is the telephone, whether landline or mobile. The key advantage of a conversation is that problems and queries can be immediately sorted out. However the biggest problem is that there is no record of any decisions that have been made. It is therefore often wise to ask for written confirmation of anything that has been agreed, perhaps in the form of an email.

Construction is one of the many industries that still prefer to have hard copies of documents. It has been made much easier to send copies of documents as email attachments. The problem though is having an available printer of sufficient quality and size to print off attached documents.

Performance reviews

As you progress through your construction career you will be expected to attend performance reviews. This is another form of communication between you and your immediate supervisor. Certain levels of performance will be expected and will have been agreed at previous reviews. At each review your performance, compared to those standards, will be examined. It gives both sides an opportunity to look at progress. It can help identify areas where you might need additional training or support. It may also show areas of your work that need improvement and more effort from you.

Meetings

Meetings also offer important opportunities for communication. They are usually quite structured and will have a series of topics that form what is known as an **agenda**.

Meetings should give everyone the opportunity to contribute and make suggestions as to how to go forward on particular projects and deal with problems. Individuals are often given the job of preparing information for meetings and then presenting it for discussion. A disadvantage of meetings is that while they are happening construction work is not taking place. This means that it is important to run meetings efficiently and not waste time – but also to ensure that everything that needs to be discussed is covered so that extra meetings do not have to be arranged.

Letters

Today emails have largely overtaken more traditional forms of communication, but letters can still be important. Letters obviously need to be delivered so take longer to arrive than emails but sometimes things do need to be sent through the post. It is polite to put in a **covering letter** with documents or other written communication with clients.

Signs and posters

On a daily basis you will also see a range of signs and posters around larger construction sites. Signs are used to communicate either warnings or information and a full list of different types of sign, particularly those relating to health and safety, can be seen in Chapter 1. Their purpose is to be clear and informative. Posters are often put up in communal areas, such as where you might have lunch or keep your personal belongings. These are designed to be simple and to give you vital information. One disadvantage of signs and posters is that they are a one-way form of communication so if you need more information about them you will need to speak to your supervisor.

REED TIP

A good supervisor will make sure you understand what is expected of you in terms of quality, quantity, the speed of the work and how you'll be working with other trades.

KEY TERMS

Agenda

– a brief list of topics to be discussed at a meeting, outlining any decisions that need to be made.

Covering letter

– this is a very brief letter, often just one paragraph long, which states the purpose of the communication and lists any other documents that have been included.

TEST YOURSELF

1. If a drawing is at a scale of 1:500, each millimetre in the drawing represents how much on the ground?

   a. 1 m

   b. 500 cm

   c. 500 mm

   d. 500 m

2. What is the other term used to describe an orthographic projection?

   a. First angle

   b. Second angle

   c. Assembly drawing

   d. Isometric

3. Which of the following are examples of a manufacturer's technical information?

   a. Data sheets

   b. User instructions

   c. Catalogues

   d. All of these

4. On a drawing, if you were to see the letters FDN, what would that mean?

   a. The signature of the architect

   b. Foundation Design Network

   c. Foundations

   d. Full distance

5. If a drawing is at a scale of 1:5, how many times smaller is the drawing than the real object?

   a. 5 times

   b. 50 times

   c. Half the size

   d. 500 times

6. Which of the following values is pi?

   a. 3.121

   b. 3.424

   c. 3.142

   d. 3.421

7. Which document is used to give detailed sets of requirements that cover the construction, features, materials and finishes?

   a. Work programme

   b. Purchase order

   c. Policy document

   d. Job specification

8. What is VAT?

   a. Volume Added Turnover

   b. Vehicle Attendance Tax

   c. Voluntary Aided Trading

   d. Value Added Tax

9. Which individual on a typical site would sign off timesheets?

   a. Architect

   b. Site manager/supervisor

   c. Delivery driver

   d. Customer

10. Which are the two main types of communication?

    a. Verbal and written

    b. Telephones and emails

    c. Meetings and memorandum

    d. Plans and faxes

Chapter 3

Unit CSA–L2Core05
UNDERSTANDING CONSTRUCTION TECHNOLOGY

LEARNING OUTCOMES

LO1: Understand the principles of foundation construction

LO2: Understand the principles of floor construction

LO3: Understand the principles of wall construction

LO4: Understand the principles of roof construction

LO5: Understand the supply of utilities and services within construction

LO6: Understand the principle of sustainability within construction

INTRODUCTION

The aim of this chapter is to:

● help you understand the range of building materials used within the construction industry

● help you understand their suitability in the construction of modern buildings.

FOUNDATION CONSTRUCTION

Foundations are the primary element of a building as they support and protect the superstructure (the visible part of the building) above. Foundations are part of the substructure of the building, meaning that they are not visible once the building has been completed.

Foundations spread the load of the superstructure and transfer it to the ground below. They provide the building with structural stability and help to protect the building from any ground movement.

Purpose of foundations

It is important to work out the necessary width of foundations. This depends on the total load of the structure and the load-bearing capacity of the ground or subsoil on which the building is being constructed. This means:

● wide foundations are used when the construction is on weak ground, or the superstructure will be heavy

● narrow foundations are used when the subsoil is capable of carrying a heavy weight, or the building is a relatively light load.

The load that is placed on the foundations spreads into the ground at 45°. **Shear failure** will take place if the thickness of the foundations (T) is less than the projection of the wall or column face on the edge of the foundations (P). This is what leads to subsidence (the ground under the structure sinking or collapsing).

As we will see in this section, the depth of the foundation is dependent on the load-bearing capacity of the subsoil. But for the most part foundations should be 200 mm to 300 mm thick.

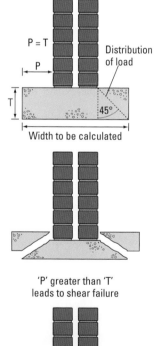

Figure 3.1 Foundation properties

Different types of foundation

The traditional **strip foundation** is quite narrow and tends to be used for low-rise buildings and dwellings. Most buildings have had unreinforced strip foundations and they were constructed with either brick or block masonry up to the damp course level. Strip foundations can be stepped on sloping ground, in order to cut down on the amount of excavation needed. In poor soil conditions, deep strip foundations can also be used.

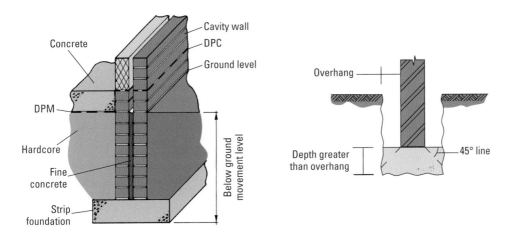

Figure 3.2 Unreinforced strip foundation

Narrow deep strip or trench fill foundations are dug to the foundation depth and then filled with concrete. This reduces excavation, as no bricks or blocks have to be laid into the trench. Trench fill:

* reduces the need to have a wide foundation

* reduces construction time

* speeds up the construction of the footings.

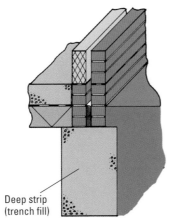

Figure 3.3 Trench fill foundations

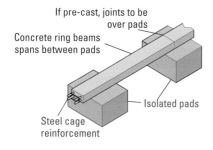

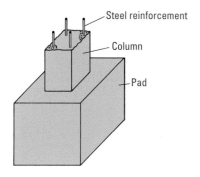

Figure 3.4 Pad foundations

Pad foundations tend to be used for structures that have either a concrete or a steel frame. The pads are placed to support the columns, which transfer the load of the building into the subsoil.

Pile foundations tend to be used for high-rise buildings or where the subsoil is unstable. Holes are bored into the ground and filled with concrete or pre-cast concrete, steel or timber posts are driven into the ground. These piles are then spanned with concrete ring beams with steel reinforcement so that the load of the building is transferred deeper into the ground below. Pile foundations can be short or long depending on how high the building is or how bad the soil conditions are.

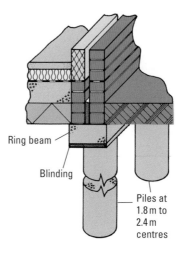

Figure 3.5 Pile foundations

Raft foundations are used when there is a danger that the subsoil is unstable. A large concrete slab reinforced with steel bars is used to outline the whole footprint of the building. It has an edge beam to take the load from the walls, which is transferred over the whole raft. This means that the building effectively 'floats' on the ground surface on top of the concrete raft.

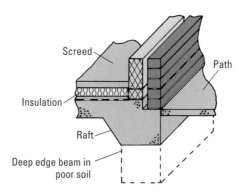

Figure 3.6 Raft foundations

Selecting a foundation

One of the first things that a structural engineer will look at when they investigate a site is the nature of the soil and issues such as the water table (where groundwater begins). The type of soil or ground conditions are be very important, as is the possibility of ground movement.

Table 3.1 shows different types of subsoil and how they can affect the choice of foundation.

Subsoil type	Characteristics
Rock	High load bearing but there may be cracks or faults in the rock, which could collapse.
Granular	Medium to high load bearing and can be compacted sand or gravel. If there is a danger of flooding the sand can be washed away.
Cohesive	Low to medium load bearing, such as clay and silt. These are relatively stable, but may have problems with water.
Organic	Low load bearing, such as peat and topsoil. Organic material must be removed before starting the foundations. There is also a great deal of air and water present in the soil.

Table 3.1 Different types of subsoil

The ground may move, particularly if the conditions are wet, extremely dry or there are extremes of temperature. Clay, for example, will shrink in the hot summer months and swell up again in the wet winter months. Frost can affect the water in the ground, causing it to expand.

Ground movement is also affected by the proximity of trees and large shrubs. They will absorb water from the soil, which can dry out the subsoil. This causes the soil underneath the foundations to collapse.

The end use of the building

The other key factor when selecting a foundation is the end use of the building:

* Strip foundation – this is the most common and cheapest type of foundation. Strip foundations are used for low to medium rise domestic and industrial buildings, as the load-bearing will not be high.

* Raft foundation – this is only ever really used when the ground on which the building is being constructed is very soft. It is also sometimes used when the ground across the area is likely to react in different ways because of the weight of the building. In areas of the UK where there has been mining, for example, raft foundations are quite common, as the building could subside. The raft is a rigid, concrete slab reinforced with steel bars. The load of the building is spread across the whole area of the raft.

* Piled foundations – these are used for high-rise buildings where the building will have a high load or where the soil is found to be poor.

Materials used in the construction of foundations

Concrete

Concrete is used to produce a strong and durable foundation. The concrete needs to be poured into the foundation with some care. The size of the foundation will usually determine whether the concrete is actually mixed on site or brought in, in a ready-mixed state, from a supplier. For smaller foundations a concrete mixer and wheelbarrows are usually sufficient. The concrete is then poured into the foundation using a chute.

Concrete consists of both fine and coarse aggregate, along with water, cement and additives if required.

Aggregates

Aggregates are basically fillers. The coarse aggregate is usually either crushed rock or gravel. The grains are 5 mm or larger.

Fine aggregate is usually sand that has grains smaller than 5 mm.

The fine aggregate fills up any gaps between the particles in the coarse aggregate.

Cement

Cement is an adhesive or binder. It is Portland stone, crushed, burnt and crushed again and mixed with limestone. The materials are powdered and then mixed together to create a fine powder, which is then fired in a kiln.

Water

Potable water, which is water that is suitable for drinking, should be used when making concrete. The reason for this is that drinkable water has not been contaminated and it does not have organic material in it that could rot and cause the concrete to crack. The water mixes with the cement and then coats the aggregate. This effectively bonds everything together.

Additives

Additives, or admixtures, make it possible to control the setting time and other aspects of fresh concrete. It allows you to have greater control over the concrete. Common add mixtures can accelerate the setting time, or reduce the amount of water required. They can:

* give you higher strength concrete

* provide protection against corrosion

* accelerate the time the concrete needs to set

* reduce the speed at which the concrete sets

* provide protection against cracking as the concrete sets (prevent shrinkage)

* improve the flow of the concrete

* improve the finish of the concrete

* provide hot or cold weather protection (a drop or rise in temperature can change the amount of time that a concrete needs to set, so these add mixtures compensate for that).

Reinforcement

Steel bars or mesh can be used to give the foundation additional strength and support. It can also stop the foundation from cracking. Concrete is very good at dealing with loads, so weight coming from above is something concrete can deal with. But when concrete foundations are wide, and parts of them are under tension, there is a danger it may crack.

Concrete should also be levelled, usually with a vibrator or a compactor, although newer types of concrete are self-compacting. All concrete needs to be laid on well-compacted ground.

Figure 3.7 Reinforcement using steel bars (or mesh)

FLOOR CONSTRUCTION

For most domestic buildings floorboards or sheets are laid over timber joists. In other cases, and in most industrial buildings, the ground floors have a block and beam construction with hard core. They then have a damp-proof membrane and over the top is solid concrete.

A floor is a level surface that provides some insulation and carries any loads (for example furniture) and to transfer those loads.

The ground floors have additional purposes. They must stop moisture from entering the building from the ground. They also need to prevent plant or tree roots from entering the building.

Ground floors

For ground floors there are two options:

* Solid – is in contact with the ground.

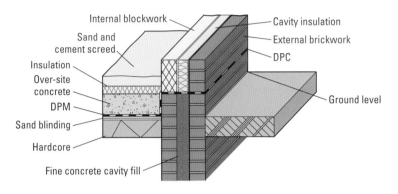

Figure 3.8 Solid ground floors

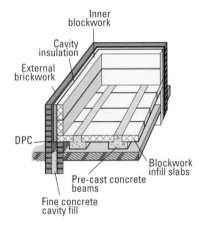

Figure 3.9 Suspended ground floors

* Suspended – the floor does not touch the ground and spans between walls in the building. Effectively there is a void beneath the floor, with air bricks in external walls to allow for ventilation.

The options for ground floors are more complicated than those for upper floors. This is because the ground floors need to perform several functions. It is quite rare for modern buildings to have timber joists and floorboards. Suspended ground floors and traditional timber floors tend to be seen in older buildings. It is far more common to have solid ground floors, or to have timber floors over concrete floors, which are known as floating ground floors.

The key options are outlined in Table 3.2.

Type of floor	Construction and characteristics
Solid	One construction method is to use hard core as the base, with a layer of sand and a layer of insulation such as Celotex, usually 100 mm thick, and then covered with a damp-proof membrane. The concrete is then poured into the foundation. To provide a smooth finish for floor finishes a cement and sand screed is applied, usually after the building has been made watertight..
Timber suspended	A similar process to a solid ground floor is carried out but then, on top of this, dwarf or sleeper walls are built. These are used to support the timber floor. Air bricks are also added to provide necessary ventilation. Joists are then spaced out along the dwarf walls. A damp-proof course is inserted under the floor joists and then floorboards or sheets placed on top of the joists.
Beam and block suspended	Concrete beams and lightweight concrete slabs or blocks are used to create the basic flooring. The beams are evenly spaced across the foundation and gaps between the beams are filled with blocks to form the floor. The blocks and beams are then insulated and it is finished off with either a cement screed or a timber floating floor.
Floating	This timber construction goes over the top of concrete floors. Bearers are put down and then the boarding or sheets are fixed to the bearers. The weight of the boards themselves hold them in place.

Table 3.2 Construction of ground floors

Upper floors

Timber is usually used for these suspended floors in homes and other types of dwellings. In industrial buildings concrete tends to be used.

Timber suspended upper floor

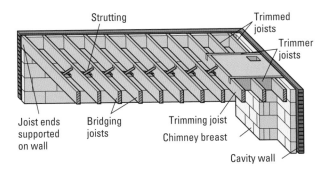

Strutting

Trimmed joists

Trimmer joists

Joist ends supported on wall

Bridging joists

Trimming joist

Chimney breast

Cavity wall

Concrete suspended upper floor

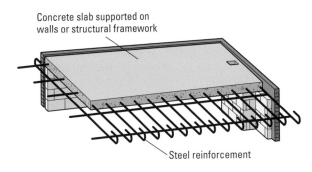

Concrete slab supported on walls or structural framework

Steel reinforcement

Figure 3.10 Upper floors

For dwellings, bridging joists are used. These are supported at their ends by load-bearing walls. Boarding or sheets provide the flooring for the room on the top of the joists. Underneath the joists plasterboard creates the basis of the ceiling for the room below.

It is also possible to fill the voids between the floorboards and the plasterboard with insulation. Insulation not only helps to prevent heat loss, but can also reduce noise.

Concrete suspended floors are usually either cast on site or available as ready-cast units. They are effectively locked into the structure of the building by steel reinforcement. If the concrete floors are being cast on site then **formwork** is needed. Concrete floors tend to be used in many modern buildings, particularly industrial ones, as they offer greater load bearing capacity, have greater fire resistance and are more sound resistant.

KEY TERMS

Formwork

– can also be known as shuttering. It is a temporary structure or mould that supports and shapes wet concrete until it cures and is able to be self-supporting.

WALL CONSTRUCTION

Walls have a number of different purposes:

* They hold up the roof.

* They provide protection against the elements.

* They keep the occupants of the building warm.

Many buildings now have double walls, which means as follows:

* The outside wall is a wet one because it is exposed to the elements outside the building.

* The internal wall is dry but it needs to be kept separate from the outside wall by a cavity.

* The cavity or gap acts as a barrier against damp and also provides some heat insulation.

* The cavity can be completely filled or part-filled depending on the insulation value required by Building Regulations.

Within the building there are other walls. These internal walls divide up the space within the building. These do not have to cope with all of the demands of the external walls. As a result, they do not necessarily have to be insulated and are, therefore, thinner. They are block and then covered with plaster. Alternatively they can be a timber framework, which is also known as stud work, and again can be covered with plasterboard.

Different types of wall construction and structural considerations

In addition to walls being external or internal, they can also be classed as being load bearing or non-load bearing.

Internal walls can be either load bearing or non-load bearing. In both internal and external walls, where they are load bearing, any gaps or openings for windows or doors have to be bridged. This is achieved by using either arches or lintels. These support the weight of the wall above the opening.

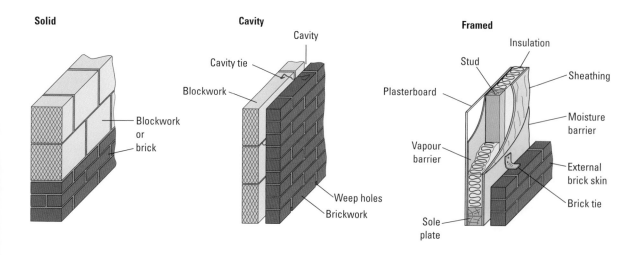

Figure 3.11 Some examples of external wall construction

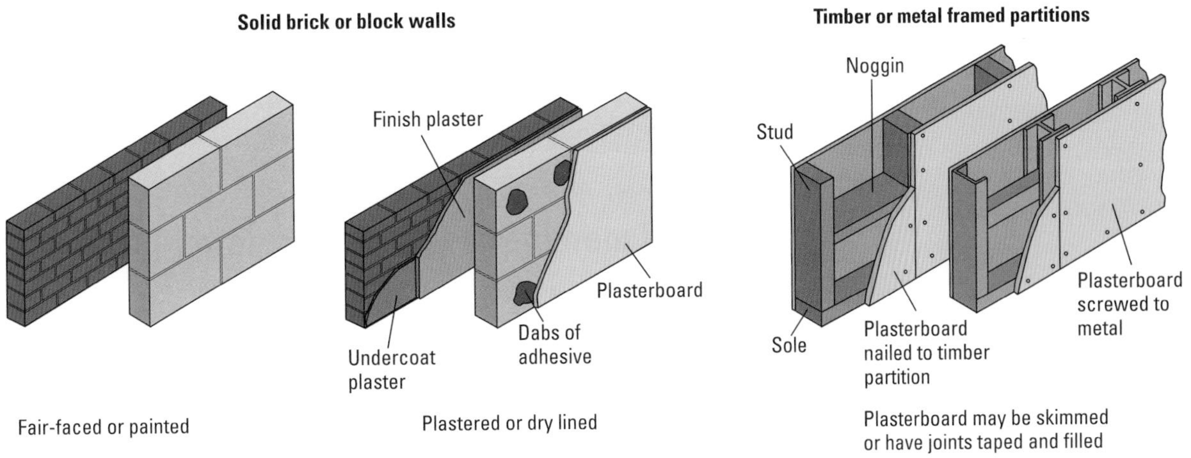

Solid brick or block walls

Timber or metal framed partitions

Finish plaster

Plasterboard

Undercoat plaster

Dabs of adhesive

Noggin

Stud

Sole

Plasterboard nailed to timber partition

Plasterboard screwed to metal

Fair-faced or painted

Plastered or dry lined

Plasterboard may be skimmed or have joints taped and filled

Figure 3.12 Some examples of internal wall construction

Solid masonry

In modern builds solid masonry is quite rare, as it uses up a lot of bricks and blocks. External solid walls tend to be much thinner and made from lightweight blocks in modern builds. They will have some kind of waterproof surface over the top of them, which can be made of render, plastic, metal or timber.

Cavity masonry

As we have seen, cavity walls have an outer and an inner wall and a cavity between them. Usually solid walling, or blockwork, is built up to ground level and then the cavity walling continues to the full height of the building. These cavity walls are ideal for most buildings up to medium height.

Many industrial buildings have cavity walls for the lower part of the building and then have insulated steel panels for the top part of the building.

The usual technique is to have brick for the outer wall and an insulating block for the inner wall. The gap or cavity can then be filled with an insulation material.

Timber framed

Panels made of timber, or in some cases steel, are used to construct walls. They can either be load bearing or non-load bearing and can also be used for the outside of the building or for internal walling. The panels are solid structures and the spaces between the vertical struts (studs) and the horizontal struts (head or sole plates) can be filled with insulation material.

Internal walls or partitions

Internal walls tend to be either solid or framed. Solid walls are made up from blocks. In many industrial buildings the blocks are actually exposed and can be left in their natural state or painted. In domestic buildings plasterboard is usually bonded to the surface and then plastered over to provide a smoother finish.

It is more common for domestic buildings to have framed internal walls, which are known as stud partitions. These are exactly the same as other framed walling, but will usually have plasterboard fixed to them. They would then receive a skimmed coat of plaster to provide the smooth finish.

Damp-proof membrane (DPM) and damp-proof course (DPC)

Damp-proof membranes are installed under the concrete in ground floors in order to ensure that ground moisture does not enter the building. Effectively the membrane waterproofs the building.

Damp-proof courses are a continuation of the damp-proof membrane. They are built into a horizontal course of either block and brickwork, which is a minimum of 150mm above the exterior ground level. DPCs are also designed to stop moisture from coming up from the ground, entering the wall and then getting into the building. The most common DPC is a polythene sheet called visqueen DPC. It comes in rolls to the appropriate width for the wall.

In older buildings lead, bitumen or slate would have been used as a DPC.

ROOF CONSTRUCTION

In a country such as the UK, with a great deal of rain and sometimes snowy weather, it makes sense for roofs to be pitched. Pitched means built at an angle. The idea is that the rain and snow falls down the angle and off the edge of the roof or into gutters rather than lying on the roof.

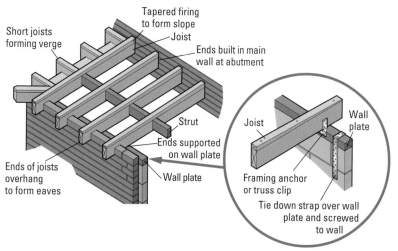

Short joists forming verge

Tapered firing to form slope

Joist

Ends built in main wall at abutment

Strut

Ends supported on wall plate

Ends of joists overhang to form eaves

Wall plate

Joist

Framing anchor or truss clip

Wall plate

Tie down strap over wall plate and screwed to wall

Figure 3.13 Flat roof structure

However, not all roofs are pitched. In fact many domestic dwelling extensions have flat roofs. A great number of industrial buildings have entirely flat roofs. The problem with a flat roof is that it needs to be able to support itself, but just as importantly it needs to be able to carry the additional weight of snow or rain. This means that large flat roofs may have to have steel sections (known as trusses) or even reinforced concrete and beams to increase their load-bearing capacity.

Roofs also provide stability to the walls by tying them together. As we will see, there are several different types of roof. These are usually identified by their pitch or shape.

Types of roof construction

The roof is made up of the rafters and beams. Everything above the framework is regarded as a roof covering, such as slates, tiles and felt.

Table 3.3 outlines some of the key characteristics of different types of roof.

Roof type	Characteristics	How it looks
Flat	This is a roof that has a slope of less than 10°. Generally flat roofs are used for smaller extensions to dwellings and on garages. Traditionally they would have had bitumen felt, although it is becoming more common for fibreglass to be used.	 Figure 3.14
Mono-pitch	This is a roof that has a single sloping surface but is not fixed to another building or wall. The front and back walls could be different heights, or the other exposed surface of the roof is **perpendicular**.	 Figure 3.15
Gambrel roof	This is a roof that has two differently angled slopes. Usually the upper part of the roof has a fairly shallow pitch or slope and the lower part of the roof has a steeper slope.	 Figure 3.16
Couple roof	This is often called gable end and is one of the most common types of roof for dwellings. A gable is a wall with a triangular upper part. This supports the roof in construction using purlins. This means that the roof has two sloping surfaces, which come down from the ridge to the eaves.	 Figure 3.17
Hipped roof	Hipped roofs have slopes on three or four four sides. There are also hipped roofs with single, straight gables.	 Figure 3.18
Lean-to	A lean to is similar to a mono-pitched roof except it is abutted to a wall. The slope is greater than 10°. The higher part of the roof is fixed to a higher wall.	 Figure 3.19

Table 3.3 Different types of roof

Roofing components

Each visible part of a roof has a specific name and purpose. Table 3.4 explains each of these individual features.

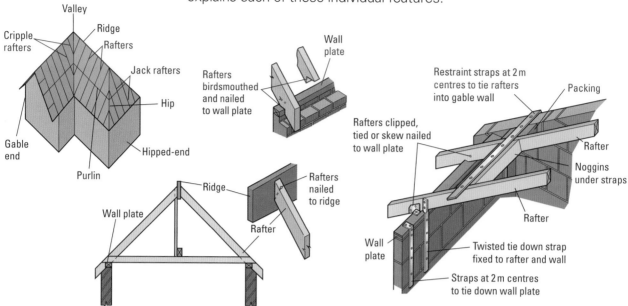

Figure 3.20 Traditional cut roof details

Roof feature	Description
Ridge	This is the top of the roof and the junction of the sloping sides. It is the peak, where the rafters meet.
Purlin	This is a beam that supports the mid-span section of rafters.
Firings	These are angled pieces of timber that are placed on the rafters to create a slope.
Batten	Roof battens are thin strips, usually of wood, which provide a fixing point for either roofing sheets or roof tiles.
Tile	These can be made from clay, slate, concrete or plastic. They are placed in regular, overlapping rows and fixed to the battens.
Fascia	This is a horizontal, decorative board. It is usually a wooden board, although it can be PVC. It is fixed to the ends of the rafters at eaves level and is both a decorative feature and a fixing for rainwater goods.
Wall plate	This is a horizontal timber that is placed at the top of a wall at eaves level. It holds the ends of joists or rafters.
Bracings	Roof rafters need to be braced to make them more rigid and stable. These bracings prevent the roof from buckling. Usually there are several braces in a typical roof.
Felt	Roofing felt has two elements – it has a waterproofing agent (bitumen) and what is known as a carrier. The carrier can be either a polyester sheet or a glass fibre sheet. Roofing felt tends to be used for flat roofs and for roofs with a shallow pitch.
Slate	Slate roofing tiles are usually fixed to timber battens with double nails. They have a lifespan of between 80 and 100 years.
Flashings	Wherever there is a joint or angle on a roof, a thin sheet of either lead or another waterproof material is added. In the past this tended always to be lead. Many different types of flashing can now be used but all have the role of preventing water penetrating into joints, such as on abutments to walls and around the chimney stack.
Rafter	Roof rafters are the main structural components of the roof. They are the framework. They rest on supporting walls. The rafters are set at an angle on sloped roofs or horizontal on a flat roof.
Apex	The apex is the highest point of the roof, usually the ridge line.

Roof feature	Description
Soffit	Soffits are the lower part, or overhanging part, of the eaves. In other words they are the underside of the eaves. A flat section of timber or plastic is usually fixed to the soffit to ensure water tightness.
Bargeboard	This is an ornamental feature, which is fixed to the gable end of a roof in order to hide the ends of roof timbers.
Eaves	These are the area found at the foot of the rafter. They are not always visible as they can be flush. In modern construction, the eaves have two parts: the visible eaves projection and the hidden eaves projection.

Table 3.4 Parts of a roof

Roof coverings

There are many different types of materials that can be used to cover the roof. Even tiles and slates come in a wide variety of shapes and sizes, along with colours and different finishes.

In many cases the type of roof covering is determined by the traditional and local styles in the area. Local authorities want roof coverings that are not too far from the common style in the area. This does not stop manufacturers from coming up with new ideas, however, which can add benefits during construction. There is much innovation and labour saving that also helps to minimise build costs.

Affordable clay tiles, for example, make it possible to use traditional materials that had been out of the budget of many construction jobs for a number of years.

The Table 3.5 outlines some of the more common types of roof covering and describes their main characteristics and use.

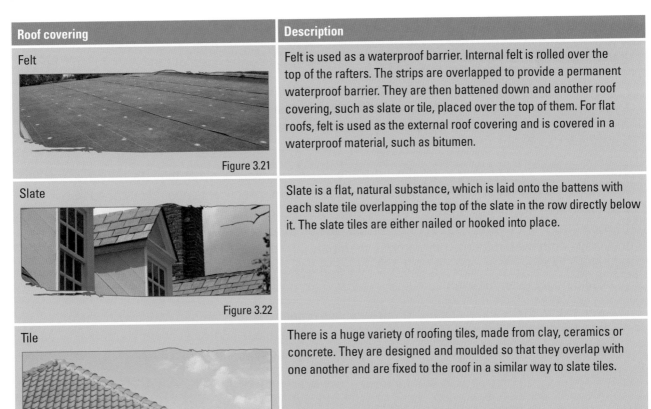

Roof covering	Description
Felt Figure 3.21	Felt is used as a waterproof barrier. Internal felt is rolled over the top of the rafters. The strips are overlapped to provide a permanent waterproof barrier. They are then battened down and another roof covering, such as slate or tile, placed over the top of them. For flat roofs, felt is used as the external roof covering and is covered in a waterproof material, such as bitumen.
Slate Figure 3.22	Slate is a flat, natural substance, which is laid onto the battens with each slate tile overlapping the top of the slate in the row directly below it. The slate tiles are either nailed or hooked into place.
Tile Figure 3.23	There is a huge variety of roofing tiles, made from clay, ceramics or concrete. They are designed and moulded so that they overlap with one another and are fixed to the roof in a similar way to slate tiles.

Roof covering	Description
Metals Figure 3.24	There are many different types of metal roof covering, such as corrugated sheets, flat sheets, box profile sheets or even sheets that have a tile effect. The metal is galvanised and plastic coated to provide a durable and long-lasting waterproof surface.

Table 3.5 Roof coverings

CASE STUDY

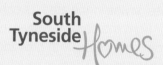

South Tyneside Homes

South Tyneside Council's
Housing Company

How to impress in interviews

Andrea Dickson and Gillian Jenkins sit on the interview panels for apprenticeship applications at South Tyneside Homes.

'Interviews are all about the three Ps: Preparation, Presentation and Personality.

An applicant should turn up with some knowledge about the apprenticeship programme and the company itself. For example, knowing how long it is, that they have to go to college and to work – and don't say, "I was hoping you'd tell me about it"! If they've done a bit of research, it will show through and work in their favour – especially if they can explain why it is that they want to work here.

It sets them up for the interview if they come in smartly dressed. We're not marking them on that, but it does show respect for the situation. It's still a formal process and although we try to make them feel at ease as much as we possibly can, there's no getting away from the fact that they're applying for a job and it is a formal setting.

The interviews are a chance to tell the company about themselves: what they do in their spare time, what their greatest achievements have been and why. Applicants should talk about what interests them, for example, are they really interested in becoming a joiner or is that something their parents want them to do? An apprenticeship has to be something they really want to do – if they have enthusiasm for the programme, then they'll fly through it. If not, it's a very long three to four years. Without that passion for it, the whole process will be a struggle; they'll come in late to work and might even fail exams.

We also talk to them about any customer service experiences they've had, working in a team, project working (for example, a time you had to complete a task and what steps you took), as well as asking some questions about health and safety awareness.'

SUPPLY OF UTILITIES AND SERVICES

Most but not all dwellings and other structures are connected in some way to a wide range of utilities and services. In the majority of cities, towns and villages structures are connected to key utilities and services, such as a sewer system, potable (drinking) water, gas and electricity. This is not always the case for more remote structures.

Whenever construction work is carried out, whether it is on an existing structure or a new build, the supply of utilities and services or the linking up of these parts of the **infrastructure** are very important. Often they will require the services of specialist engineers from the **service provider**.

Table 3.6 outlines the main utilities and services that are provided to most structures.

Utility or service	Description
Drainage	Drainage is delivered by a range of water and sewerage companies. They are responsible for ensuring that surface water can drain away into their system.
Waste water and sewerage	Any waste water and sewage generated by the occupants of a structure needs to have the necessary pipework to link it to the main sewerage system. It is then sent to a sewage treatment works via the pipework. Remote areas may not be connected to the sewerage system so use septic tanks and cesspools.
Water	Each structure should be linked to the water supply that provides wholesome, potable drinking water. The pipework linking the structure to the water supply needs to be protected to ensure that backflow from any other source does not contaminate the system.
Gas	Each area has a range of different gas suppliers. This is delivered via a service pipe from the main system into the structure. Areas that do not have access to the main gas supply system use gas contained in bottles.
Electricity	The National Grid provides electricity to a variety of different electricity suppliers. It is the National Grid that operates and maintains the cabling. There are around 28 million individual customers in the UK.
Communications (telephone, data, cable)	There are several ways in which telecommunications can be linked to a structure. Traditional telephone poles hold up copper cables and not only provide telephone but also internet access to structures. In cities and many of the larger towns this system is being replaced by cables that are fibre optic and run underground. These are then linked to each individual structure.
Ducting (heating and ventilation)	Heating and ventilation engineers install and maintain duct work. The complex systems are known as **HVAC**. These systems can transfer air for heating or cooling of the structure. The overall system can also provide hot and cold water systems, along with ventilation.

Table 3.6 Services and utilities

SUSTAINABILITY AND INCORPORATING SUSTAINABILITY INTO CONSTRUCTION PROJECTS

Sustainability is something that we all need to be concerned about as the earth's resources are used up rapidly and climate change becomes an ever-bigger issue. Carbon is present in all fossil fuels, such as coal or natural gas. Burning fossil fuels releases carbon dioxide, which is a greenhouse gas linked to climate change.

Energy conservation aims to reduce the amount of carbon dioxide in the atmosphere. The idea is to do this by making buildings better insulated and, at the same time, make heating appliances more efficient. Sustainability also means attempting to generate energy using renewable and/or low or zero carbon methods.

According to the government's Environment Agency, sustainable construction means using resources in the most efficient way. It also means cutting down on waste on site and reducing the amount of materials that have to be disposed of and put into **landfill**.

In order to achieve sustainable construction the Environment Agency recommends

* reducing construction, demolition and excavation waste that needs to go to landfill

* cutting back on carbon emissions from construction transport and machinery

* responsibly sourcing materials

* cutting back on the amount of water that is wasted

* making sure construction does not have a negative impact on **biodiversity**.

Sustainable construction and incorporating it into construction projects

In the past buildings were generally constructed as quickly as possible and at the lowest cost. More recently the idea of sustainable construction focuses on ensuring that the building is not only of good quality and that it is affordable, but that it is also efficient in terms of energy use and resources.

Sustainable construction also means having the least negative environmental impact. So this means minimising the use of raw materials, energy, land and water. This is not only during the build but also for the lifetime of the building.

Finite and renewable resources

We all know that resources such as coal and oil will eventually run out. These are examples of finite resources.

Oil, however, is not just used as fuel – it is in plastic, dyes, lubricants and textiles. All of these are used in the construction process.

Renewable resources are those that are produced either by moving water, the sun or the wind. Materials that come from plants, such as biodiesel, or the oils used to make adhesives, are all examples of renewable resources.

Figure 3.25 Most modern new-builds follow sustainable principles

The construction process itself is only part of the problem. It is important to consider the longer-term impact and demands that the building will have on the environment. This is why there has been a drive towards sustainable homes and there is a Code for Sustainable Homes.

Construction and the environment

In 2010, construction, demolition and excavation produced 20 million tonnes of waste that had to go into landfill. The construction industry is also responsible for most illegal fly tipping (illegally dumping waste). In any year the Environment Agency responds to around 350 pollution incidents caused as a result of construction.

Regardless of the size of the construction job, everyone in construction is responsible for the impact they have on the environment. Good site layout, planning and management can help to reduce these problems.

Sustainable construction helps to encourage this because it means managing resources in a more efficient way, reducing waste, recycling where possible and reducing your **carbon footprint**.

Architecture and design

The Code for Sustainable Homes Rating Scheme was introduced in 2007. Many local authorities have instructed their planning departments to encourage sustainable development. This begins with the work of the architect who designs the building.

DID YOU KNOW?

Search on the internet for 'sustainable building' and 'improving energy efficiency' to find out more about the latest technologies and products.

KEY TERMS

Carbon footprint

– This is the amount of carbon dioxide produced by a project. This not only includes burning carbon-based fuels such as petrol, gas, oil or coal, but includes the carbon that is generated in the production of materials and equipment.

Local authorities ask that architects and building designers:

* ensure the land is safe for development – so if it is contaminated this is dealt with first

* ensure access to and protect the natural environment – this supports biodiversity and tries to create open spaces for local people

* reduce the negative impact on the local environment – buildings should keep noise, air, light and water pollution down to a minimum

* conserve natural resources and cut back carbon emissions – this covers energy, materials and water

* ensure comfort and security – good access, close to public transport, safe parking and protection against flooding.

Figure 3.26 Sustainable developments aim to be pleasant places to live

Using locally managed resources

The construction industry imports nearly 6 million cubic metres of sawn wood each year. However there is plenty of scope to use the many millions of cubic metres of timber produced in managed forests, particularly in Scotland.

Local timber can be used for a wide variety of different construction projects:

* Softwood – including pines, firs, larch and spruce – for panels, decking, fencing and internal flooring.

* Hardwood – including oak, chestnut, ash, beech and sycamore – for a wide variety of internal joinery.

Eco-friendly, sustainable manufactured products and environmentally resourced timber

There are now many suppliers that offer sustainable building materials as a green alternative. Tiles, for example, are now made from recycled plastic bottles and stone particles.

There is a National Green Specification database of all environmentally friendly building materials. This provides a checklist where it is possible to compare specifications of sustainable products to traditionally manufactured products, such as bricks.

Simple changes can be made, such as using timber or ethylene-based plastics instead of uPVC window frames, to ensure a building uses more sustainable materials.

As we have seen, finding locally managed resources such as timber makes sense in terms of cost and in terms of protecting the environment. There are many alternatives to traditional resources that could help protect the environment.

The Timber Trade Federation produces a Timber Certification System. This ensures that wood products are labelled to show that they are produced in sustainable forests.

Around 80 per cent of all the softwood used in construction comes from Scandinavia or Russia. Another 15 per cent comes from the rest of Europe, or even North America. The remaining 5 per cent comes from tropical countries, and is usually sourced from sustainable forests.

Figure 3.27 Window frames made from timber

DID YOU KNOW?

www.recycledproducts. org.uk has a long list of recycled surfacing products, such as tiles, recycled wood and paving and details of local suppliers.

Alternative methods of building

The most common type of construction is, of course, brick and block work. However there are plenty of other options:

* Timber frame

* Insulated concrete formwork – where a polystyrene mould is filled with reinforced concrete.

* Structural insulated panels – where buildings are made up of rigid building boards, rather like huge sandwiches.

* Modular construction – this uses similar materials and techniques to standard construction, but the units are built off site and transported ready-constructed to their location.

Figure 3.28 Timber Certification System

Figure 3.29 Green roofing

Figure 3.30 Flooring made from cork

Alternatives to roofing and flooring

There are alternatives to traditional flooring and roofing, all of which are greener and more sustainable. Green roofing has become an increasing trend in recent years. Metal roofs made of steel, aluminium or copper are lightweight and often use a high percentage of recycled metal. Solar roof shingles, or solar roof laminates, while expensive, help to reduce the use of electricity and heating of the dwelling. Some buildings even have a green roof, which consists of a waterproof membrane, a growing medium and plants such as grass or sedum.

Just as roofs are becoming greener, so too are the options for flooring. The use of bamboo, eucalyptus or cork is becoming more common. A new version of linoleum has been developed with **biodegradable**, **organic** ingredients. Some buildings are also using concrete rather than traditional timber floorboards and joists. The concrete can be coloured, stained or patterned.

An increasing trend has been for what is known as off-site manufacture (OSM). European businesses, particularly those in Germany, have built over 100,000 houses. The entire house is manufactured in a factory and then assembled on site. Walls, floors, roofs, windows and doors with built-in electrics and plumbing, all arrive on a lorry. Some manufacturers even offer completely finished dwellings, including carpets and curtains. Many of these modular buildings are actually designed to be far more energy efficient than traditional brick and block constructions. Many come ready fitted with heat pumps, solar panels and triple-glazed windows.

KEY TERMS

Biodegradable

– the material will more easily break down when it is no longer needed. This breaking down process is done by micro-organisms.

Organic

– this is a natural substance, usually extracted from plants.

Energy efficiency and incorporating it into construction projects

Energy efficiency involves using less energy to provide the same level of output. The plan is to try to cut the world's energy needs by 30 per cent before 2050. This means producing more energy efficient buildings. It also means using energy efficient methods to produce materials and resources needed to construct buildings.

Building Regulations

In terms of energy conservation, the most important UK law is the Building Regulations 2010, particularly Part L. The Building Regulations:

* list the minimum efficiency requirements

* provide guidance on compliance, the main testing methods, installation and control

* cover both new dwellings and existing dwellings.

A key part of the regulations is the Standard Assessment Procedure (SAP), which measures or estimates the energy efficiency performance of buildings.

Local planning authorities also now require that all new developments generate at least 10 per cent of their energy from renewable sources. This means that each new project has to be assessed one at a time.

Energy conservation

By law, each local authority is required to reduce carbon dioxide emissions and to encourage the conservation of energy. This means that everyone has a responsibility in some way to conserve energy:

* Clients, along with building designers, are required to include energy efficient technology in the build.

* Contractors and sub-contractors have to follow these design guidelines. They also need to play a role in conserving energy and resources when actually working on site.

* Suppliers of products are required by law to provide information on energy consumption.

In addition, new energy efficiency schemes and building regulations cover the energy performance of buildings. Each new build is required to have an Energy Performance Certificate. This rates a building's energy efficiency from A (which is very efficient) to G (which is least efficient).

Some building designers have also begun to adopt other voluntary ways of attempting to protect the environment. These include BREEAM, which is an environmental assessment method, and the Code for Sustainable Homes, which is a certification of sustainability for new builds.

**energy®
saving
trust**

Figure 3.31 The Energy Saving Trust encourages builders to use less wasteful building techniques and more energy efficient construction

DID YOU KNOW?

The Energy Saving Trust has lots of information about construction. Its website is www. energysavingtrust.org.uk.

High, low and zero carbon

When we look at energy sources, we consider their environmental impact in terms of how much carbon dioxide they release. Accordingly, energy sources can be split into three different groups:

* high carbon – those that release a lot of carbon dioxide

* low carbon – those that release some carbon dioxide

* zero carbon – those that do not release any carbon dioxide.

Some examples of high carbon, low carbon and zero carbon energy sources are given in Table 3.7 below.

High carbon energy source	Description
Natural gas or LPG	Piped natural gas or liquid petroleum gas stored in bottles
Fuel oils	Domestic fuel oil, such as diesel
Solid fuels	Coal, coke and peat
Electricity	Generated from non-renewable sources, such as coal-fired power stations
Low carbon energy source	
Solar thermal	Panels used to capture energy from the sun to heat water
Solid fuel	Biomass such as logs, wood chips and pellets
Hydrogen fuel cells	Converts chemical energy into electrical energy
Heat pumps	Devices that convert low temperature heat into higher temperature heat
Combined heat and power (CHP)	Generates electricity as well as heat for water and space heating
Combined cooling, heat and power (CCHP)	A variation on CHP that also provides a basic air conditioning system
Zero carbon energy	
Electricity/wind	Uses natural wind resources to generate electrical energy
Electricity/tidal	Uses wave power to generate electrical energy
Hydroelectric	Uses the natural flow of rivers and streams to generate electrical energy
Solar photovoltaic	Uses solar cells to convert light energy from the sun into electricity

Table 3.7 High, low and zero carbon energy sources

It is important to try to conserve non-renewable energy so that there will be sufficient fuel for the future. The fuel has to last as long as is necessary to completely replace it with renewable sources, such as wind or solar energy.

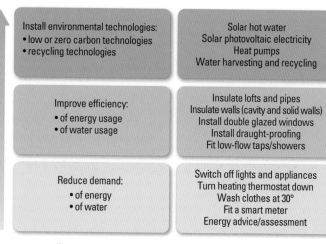

Figure 3.32 Working towards reducing carbon emissions

Alternative energy sources

There are several new ways in which we can harness the power of water, the sun and the wind to provide us with new heating sources. All of these systems are considered to be far more energy efficient than traditional heating systems, which rely on gas, oil, electricity or other fossil fuels.

Solar thermal

At the heart of this system is the solar collector, which is often referred to as a solar panel. The idea is that the collector absorbs energy from the sun, which is then converted into heat. This heat is then applied to the system's heat transfer fluid.

The system uses a differential temperature controller (DTC) that controls the system's circulating pump when solar energy is available and there is a demand for water to be heated.

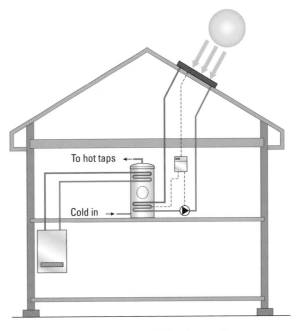

Figure 3.33 Solar thermal hot water system

In the UK, due to the lack of guaranteed solar energy, solar thermal hot water systems often have an auxiliary heat source, such as an immersion heater.

Biomass (solid fuel)

Biomass stoves burn either pellets or logs. Some have integrated hoppers that transfer pellets to the burner. Biomass boilers are available for pellets, woodchips or logs. Most of them have automated systems to clean the heat exchanger surfaces. They can provide heat for domestic hot water and space heating.

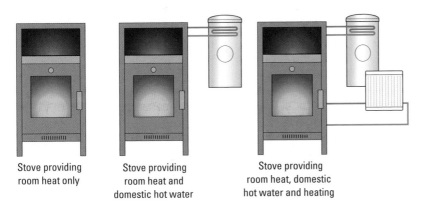

Stove providing room heat only

Stove providing room heat and domestic hot water

Stove providing room heat, domestic hot water and heating

Figure 3.34 Biomass stoves output options

Heat pumps

Heat pumps convert low temperature heat from air, ground or water sources to higher temperature heat. They can be used in ducted air or piped water **heat sink** systems.

KEY TERMS

Heat sink

– this is a heat exchanger that transfers heat from one source into a fluid, such as in refrigeration, air-conditioning or the radiator in a car.

There are different arrangements for each of the three main systems:

* Air source pumps operate at temperatures down to minus 20°C.

* Ground source pumps operate on **geothermal** ground heat.

* Water source systems can be used where there is a suitable water source, such as a pond or lake.

The heat pump system's efficiency relies on the temperature difference between the heat source and the heat sink.

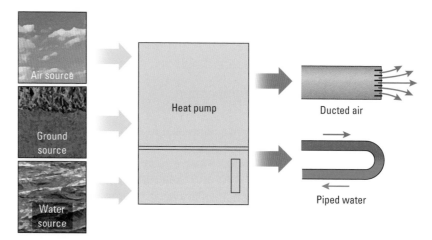

Figure 3.35 Heat pump input and output options

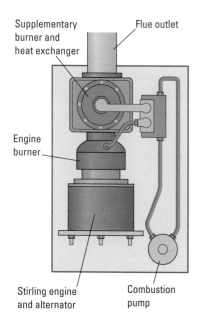

Figure 3.36 Example of a MCHP (micro combined heat and power) unit

Combined heat and power (CHP) and combined cooling heat and power (CCHP) units

These are similar to heating system boilers, but they generate electricity as well as heat for hot water or space heating (or cooling). Electricity is generated along with sufficient energy to heat water and to provide space heating.

Wind turbines

Freestanding or building-mounted wind turbines capture the energy from wind to generate electrical energy. The wind passes across rotor blades of a turbine, which causes the hub to turn. The hub is connected by a shaft to a gearbox. This increases the speed of rotation. A high speed shaft is then connected to a generator that produces the electricity.

Solar photovoltaic systems

A solar photovoltaic system uses solar cells to convert light energy from the sun into electricity.

Energy ratings

Energy rating tables are used to measure the overall efficiency of a dwelling, with rating A being the most energy efficient and rating G the least energy efficient (see Fig 3.41).

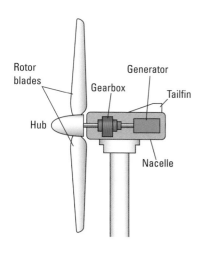

Figure 3.37 A basic horizontal axis wind turbine

Alongside this, there are environmental impact ratings (see Fig 3.40). This type of rating measures the dwelling's impact on the environment in terms of how much carbon dioxide it produces. Again, rating A is the highest, showing it has the least impact on the environment, and rating G is the lowest.

A Standard Assessment Procedure (SAP) is used to place the dwelling on the energy rating table. The ratings are used by local authorities and other groups to assess the energy efficiency of new and old housing and must be provided when houses are sold.

Preventing heat loss

Most old buildings are under-insulated and benefit from additional insulation, whether this is applied to ceilings, walls or floors.

The measurement of heat loss in a building is known as the U Value. It measures how well parts of the building transfer heat. Low U Values represent high levels of insulation. U Values are becoming more important as they form the basis of energy and carbon reduction standards.

By 2016 all new housing is expected to be Net Zero Carbon. This means that the building should not be contributing to climate change.

Many of the guidelines are now part of Building Regulations (Part L). They cover:

* insulation requirements
* openings, such as doors and windows
* solar heating and other heating
* ventilation and air-conditioning
* space heating controls
* lighting efficiency
* air tightness.

Building design

UK homes spend £2.4bn every year just on lighting. One of the ways of tackling this cost is to use energy saving lights, but also to maximise natural lighting. For the construction industry this means:

* increased window size
* orientating window angles to make the most of sunlight – south facing windows maximise sunlight in winter and limit overheating in the summer
* window design – with a variety of different types of opening to allow ventilation.

Solar tubes are another way of increasing light. These are small domes on the roof, which collect sunlight and then direct it through a tube (which is reflective). It is then directed through a diffuser in the ceiling to spread light into the room.

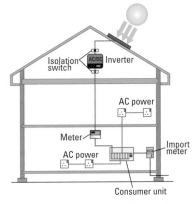

Figure 3.38 A basic solar photovoltaic system

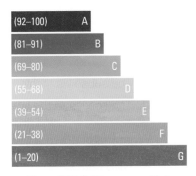

Figure 3.39 SAP energy efficiency rating table – the ranges in brackets show the percentage energy efficiency for each banding

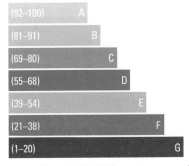

Figure 3.40 SAP environmental impact rating table

TEST YOURSELF

1. In which of the following types of building is a traditional strip foundation used?

    a. High rise

    b. Medium rise

    c. Low rise

    d. Industrial buildings

2. Which of the following is a reason for using a raft foundation?

    a. The subsoil is rock

    b. The subsoil is unstable

    c. The subsoil is stable

    d. The access to the site allows it

3. What holds down a floating floor?

    a. Nails and screws

    b. Adhesives

    c. Blocks

    d. Its own weight

4. What is another term for formwork?

    a. Shuttering

    b. Cavity

    c. Joist

    d. Boarding

5. What is the minimum distance the DPC should be above ground level?

    a. 50mm

    b. 100mm

    c. 150mm

    d. 200mm

6. A roof is said to be flat if it has a slope of less than how many degrees?

    a. 5

    b. 10

    c. 15

    d. 20

7. What shape is the upper part of a gable end?

    a. Rectangular

    b. Semi-circular

    c. Square

    d. Triangular

8. What do you call the horizontal timber that is placed at the top of a wall at eaves level in a roof, to hold the ends of joists or rafters?

    a. Fascia

    b. Bracings

    c. Wall plate

    d. Batten

9. What happens to the majority of construction demolition and excavation waste?

    a. It is buried on site

    b. It is burned

    c. It goes into landfill

    d. It is recycled

10. Which part of the Building Regulations 2010 requires construction to consider and use energy efficiently?

    a. Part B

    b. Part D

    c. Part K

    d. Part L

Chapter 4

Unit CSA L2Occ57

APPLY PLASTER MATERIALS TO INTERNAL SURFACES

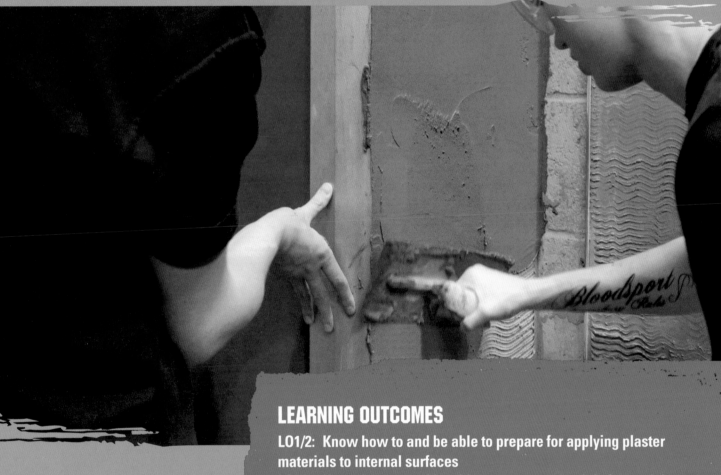

LEARNING OUTCOMES

LO1/2: Know how to and be able to prepare for applying plaster materials to internal surfaces

LO3/4: Know how to and be able to prepare materials for applying plaster materials to internal surfaces

LO5/6: Know how to and be able to apply plaster materials to internal backgrounds

INTRODUCTION

The aims of this chapter are to:

* show you how to prepare background surfaces for plastering

* help you to choose, prepare and mix the appropriate materials for plastering

* show you how to apply trims

* show you how to apply two or more coats of plaster.

Plaster is a building material used for coating walls and ceilings over a variety of background surfaces. It can be smooth or textured, and it may protect the **background surface** from damp or extreme temperatures, or it might just be decorative. Traditionally, plaster contained lime but gypsum-based plasters are now more commonly used and are easier to work with, set a lot faster and provide a high quality smooth finish. Plaster applied to an outside wall or background surface is known as **render**.

As a plasterer, you may work on sites ranging from large construction projects to renovations in clients' homes. It is likely that you will not only apply plaster materials to walls but also work with plasterboard, lay floor screeds, and produce mouldings and fix them in position. This book covers all those techniques and, over time, you may specialise in one or more of them.

Whatever type of plastering work you are doing, the quality of the end result is affected by how well you have prepared.

KEY TERMS

Plaster

– a building material used for coating wall and ceilings over a variety of background surfaces. It can be smooth or textured, and it may protect the background surface from damp or extreme temperatures, or just be decorative.

Background surface

– the surface to which plaster or render will be applied. It would normally be brickwork, blockwork or plasterboard, but may also be concrete, wood, tiles or old plaster.

Render

– the coating on an outside wall. It contains sand, cement and, in some cases, lime or additives to make it weather resistant.

Figure 4.1 A newly plastered room

PREPARING TO APPLY PLASTER MATERIALS TO INTERNAL SURFACES

Hazards associated with internal plastering

Refresh your memory about general health and safety precautions by referring back to Chapter 1.

You or your employer should always carry out a risk assessment before work begins, so that any issues with health and safety can be identified and dealt with. You must always read the risk assessment and follow its recommendations. If you see anything unsafe, report it to your supervisor at once.

Hazards when preparing backgrounds
Different backgrounds are prepared in different ways.

Dust and chemicals are hazardous substances, so ensure you wear the correct personal protective equipment (see below and Chapter 1).

Hazards when working with plaster
The main hazards when working with plaster are from the chemicals in the mix, and from the dust that arises from powdered materials. Plaster materials that may be hazardous include:

* adhesives

* cement and other gypsum-based products

* joint fillers

* lime

* mineralised meths

* polyester resins

* PVA

* shellac.

These can all damage your skin or cause illness if they are breathed in. You should make sure to do the following to reduce the risks:

* Wear additional PPE, such as overalls and a dust mask, when applying and mixing plaster.

* Store all materials in an approved, enclosed store or location.

* Refer to the manufacturers' data sheets for each substance.

* Maintain good standards of personal hygiene by washing your hands after contact with the materials and before eating, drinking and smoking.

PRACTICAL TIP

Although your manager will probably be the one to write the risk assessment, you need to be familiar with the types of hazards and controls it will cover. You might also identify risks that your manager wasn't aware of.

DID YOU KNOW?

The Health and Safety Executive (HSE) has produced guidelines for small plastering companies on drawing up risk assessments: *www.hse.gov.uk/risk/ casestudies/plasterers. htm*

Other hazards arise if, when applying plaster, you have to work at height or need to move the materials. Here are some basic controls to lessen these risks.

- If you are on a large site, a fork lift truck or pallet truck may help you to avoid carrying heavy materials, large pallets and waste.

- Transport small amounts of mixed plaster and mortar by wheelbarrow.

- When you are plastering, inspect mixers, barrows and shovels regularly for damage or defects and take them out of use for repair if faults are discovered.

- Fence off your work area, especially if you are working at height.

- If possible, use warning signs at all access points to the work area where work is being carried out above ground level.

Personal protective equipment (PPE)

The risk assessment will identify how health and safety risks should be reduced to prevent the need for PPE. Remember that PPE is always the last resort. However, it is usually still necessary to wear appropriate PPE.

Always ensure you wear the correct PPE (see Chapter 1). If you are on a construction site, you should wear the PPE provided, such as a hard hat, hi-vis jacket and safety boots. Even on smaller sites, you should wear goggles (to prevent plaster or dust from getting into your eyes).

You should always wear some kind of hand protection too. The thickness of the gloves will depend on the job, as you may need to be able to hold things, which can be difficult if your gloves are too thick. It may seem inconvenient and hinder your work but suitable thick gloves will protect your hands against cuts, abrasions and most impacts. Some plasterers prefer to wear thin latex gloves to protect their hands from chemicals but if your employer provides you with thick protective gloves, you must wear them.

Collective safety measures

If you are working at height (for example, rendering a two-storey building) there should be collective safety measures in place, like safety nets or a guardrail or scaffolding with three levels of protection:

- a hand rail at a height of between 1 m and 1.1 m

- a base board between 100 mm and 150 mm

- an intermediate rail.

Figure 4.2 Hand protection is important when plastering

Local exhaust ventilation (LEV) systems

If dust cannot be removed or controlled in any other way, your employer may require you to use LEV. This is a ventilation system that takes airborne contaminants (dusts, mists, gases, vapour or fumes) out of the air so that they cannot be breathed in.

Properly designed LEV will:

* collect the air that contains the contaminants

* make sure they are contained and taken away from people

* clean the air (if necessary) and get rid of the contaminants safely. (*Source* HSE)

The airborne contaminants are sucked into a hood, which may be small enough to be attached to a hand-held tool or big enough to walk into. You may use a different type of LEV for doing different types of plastering jobs, for example mixing plaster additives or dealing with cement dust. You should be shown how to use each type of LEV before you start working with one.

Figure 4.3 Take care when working at height

REED TIP

Employers will want to know that you understand the importance of health and safety. Make sure you know the reasons for each safe working practice.

Figure 4.4 A local exhaust ventilation system (LEV)

Interpreting information sources for applying plaster materials to internal surfaces

As we saw in Chapter 2, various types of documentation help you to plan your work and position the plaster correctly.

Working drawings are either full size or scaled, accurate illustrations of the final finish that is required. It is important to follow any supplied documentation to ensure that your work complies with Building Regulations and other requirements of the project. For example, the technical drawings may specify a particular type of plaster, such as a lime-based mix in an older building. Using the wrong type of plaster could damage the building.

You should also refer to the plaster manufacturer's technical data sheets to make sure it is the correct material for the area in which it will be applied.

Tools and equipment for internal plastering

You will need a variety of tools and equipment to prepare the background surface, and to mix and apply the plaster materials.

DID YOU KNOW?

The Control of Substances Hazardous to Health (COSHH) Regulations 2002 require these substances to be removed from the workplace.

PRACTICAL TIP

Look out for details like expanded metal lathing (EML) in narrow widths. This may be specified to cover timber lintels or timber supports on a wall.

Tools for preparing background surfaces

You will need a variety of tools and equipment in your toolbox when preparing surfaces. Table 4.1 describes the main things you will need.

Hand tool	Description and when it should be used
Bolster chisel (Fig 4.5)	Also known as a brick chisel. It has a wide, flat blade that is used to make straight cuts in hard materials like stone or brick. Those with a safety handle can be tapped with a hammer to split the material.
Broom (Fig 4.6)	A brush with a long handle for sweeping up and dusting down floors and surfaces.
Brush (Fig 4.7)	A hand-held brush can be used to apply water to the wall when testing for its suction or absorbency. A dry brush is also used for removing dust from the wall.
Crowbar (Fig 4.8)	A metal bar with a curved end that is sometimes split into two points. The hook is used to prise off plasterboard, skirting board, architraves and large chunks of plaster.
Containers (Fig 4.9)	It is useful to have several containers or buckets for different purposes. They can be used to measure or gauge the amount of material you need, as a container for mixing liquid materials (such as bonding and plaster) and for carrying materials or even tools.
Lump hammer (Fig 4.10)	Also known as a club hammer. It has a double-faced head, usually made from a heat-treated forged steel. The handle is normally made from wood, usually hickory, or synthetic resin. It may weigh between 0.5 kg and 3 kg.
Shovel (Fig 4.11)	This is for mixing materials, loading the cement mixer and for clearing rubble.
Skutch hammer (Fig 4.12)	A bricklayer's hammer with interchangeable finishing heads for trimming and tidying bricks and blocks.

Table 4.1 Hand tools and equipment used for preparing background surfaces for plastering

Figure 4.5

Figure 4.6

Figure 4.7

Figure 4.8

Figure 4.9

Figure 4.10

Figure 4.11

Figure 4.12

You will also need to use power tools – remember that these must be 110V, if you are on a construction site, and 230V if you are in a private building, such as someone's home. Never try to use a power tool if you have not been trained to do so.

Table 4.2 describes some of the power tools you may use while preparing backgrounds.

Power tool	Description and when it should be used
Hammer drill (Fig 4.13)	A drill with a hammer action to help you drill into hard material like brick. The hammer action can be turned on or off, and is usually around 10,000 BPM (blows per minute).
Industrial vacuums (Fig 4.14)	Heavy duty commercial vacuums can be used on hard and soft backgrounds, and some remove wet debris. They come in a variety of sizes, suctions and capacities. Some can be attached to drills to automatically remove dust and debris.
Rotary stripper (Fig 4.15)	The vibrating abrasive disc removes paint and other surface coatings from background surfaces.
SDS drill (Fig 4.16)	This drill is more powerful than other hammer drills and can drill through hard surfaces like stone or concrete. It can take larger drill bits (attachments), including tile removal and chisel bits.

Table 4.2 Power tools used for preparing background surfaces for plastering

Figure 4. 13

Figure 4.15

Figure 4. 14

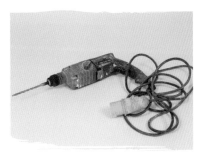

Figure 4.16

Tools for preparing and applying internal plaster materials

You will soon build up your own collection of tools, and with experience will know which tool you prefer to use for which job. Always pay as much as you can afford for good quality tools – cheap tools usually turn out to be more expensive in the long run, because they will not give such a good finish and are more likely to break quickly.

Table 4.3 describes the main tools you are likely to use when plastering.

Tool	Description
Darby	A straight-edge rule for smoothing walls and ceilings. It is made of aluminium or polyurethane and has adjustable wooden handles.
Feather edge rule (Fig 4.17)	An aluminium H-section or trapezoid straight-edged rule used to level off surfaces when floating. It can also check that the line of the wall is true and come in various lengths.
Floats (Fig 4.18)	There are several types of float, which are usually made from wood or plastic. A plasterer's float is used for flooring and rendering. A devil float has small nails in it to provide a key in the background or plaster.
Handboard/hawk (Fig 4.19)	This is a rigid aluminium, plastic or wooden square with slightly rounded corners and a handle on the base. It is for holding small amounts of plaster. You hold it in one hand and take plaster from it with a trowel held in your other hand. For this reason, a lightweight model with a padded handle is best so that it is comfortable to hold while you are working. Some types have a removable handle to make it easier to clean and so you can put it down on flat surface if you need to.
Straight edge (Fig 4.20)	Like darby and feather edge rules, these have various uses, including checking that the wall or ceiling line is true, fixing angle beads plumb with a level, checking boards are flat and true, and positioning boards when dry lining.
Spot board (Fig 4.21)	This is a square piece of plywood on legs or a stand, for holding larger amounts of material. You transfer the plaster from the spot board to the handboard / hawk when you need it.
Tin snips (Fig 4.22)	This has long handles and short blades with extra-wide jaws, and is used for cutting soft metal, such as beading or EML (see below). Straight pattern tin snips cut straight lines or gentle curves, while duckbill or Trojan-pattern snips cut larger curves and circular shapes.
Whisk (Fig 4.23)	This is a power tool with a long stem that has a whisk (paddle) attachment at the end, for mixing viscous materials like dry wall, adhesive and plaster. It will usually have a two-speed motor for mixing both thin and thick materials, as well as variable speed control. Many different types of attachment are available.

Table 4.3 Types of tools used in plastering

Figure 4.17

Figure 4.18

Figure 4.19

Figure 4.20

Figure 4.21

Figure 4.22

Figure 4.23

Trowels are very important to a plasterer. You should use the right trowel for the type of plastering you are doing. Table 4.4 describes some of the main types.

Type of trowel	Description
Bucket trowel (Fig 4.24)	This has a tapered blade leading to a square edge, to scrape plaster from the bottom of a bucket or container without causing damage.
Finishing trowel (Fig 4.25)	This is a trowel that is suitable for skimming and finishing the top coat because it has been worn in and doesn't have sharp edges that will leave a line in the plaster. Some types come pre-worn, or you can use an old floating trowel. It is made from carbon or stainless steel, although some types have a clip-on replaceable plastic blade of different sizes ranging from 300 mm × 140 mm to 700 mm × 140 mm. It is best to choose a model with a comfortable handle.
Floating/plasterer's trowel (Fig 4.26)	This is a large rectangular sheet of flat steel with a handle on the back. When it is new, it is used for backing or undercoats. When its corners become worn down it provides a smoother finish so is used as a finishing trowel for skimming. Note that some plasterers use the same trowel to apply undercoat plaster and finishing plaster.
Floor-laying trowel (Fig 4.27)	This is used for laying, mixing and spreading flooring materials like sand and cement, and for giving the screed a smooth finish. It is made from steel and is either 400 mm or 455 mm long with a pointed front.
Gauging trowel (Fig 4.28)	This is used to mix small amounts of material and to get plaster into difficult places, like corners of floor screeds. It can also be used to clean down other tools after screeding. It is made of steel and usually has a wooden or plastic handle.
Internal or external angle trowel (Fig 4.29)	This is used to get a clean finish on internal or external corner joints.

Table 4.4 Types of trowel used in plastering

Figure 4.24

Figure 4.25

Figure 4.26

Figure 4.27

Figure 4.28

Figure 4.29

You may also need ancillary equipment for working at height, such as a hop-up or stilts. Table 5.6 in Chapter 5 gives you more information about these.

PREPARING MATERIALS FOR PLASTERING INTERNAL SURFACES

The compatibility of backgrounds and applied plaster

When you are considering a particular background for plastering, you should think about these points:

* Strength: is it in good condition?

* Mechanical key: is it rough or smooth?

* Shrinkage: how will the plaster dry on it?

* Movement: are there joints to allow for the natural movement of the wall?

* Suction: what is it made of?

Different types of background surfaces have different types of **suction**. This is what determines **adhesion** – whether freshly applied plaster will stick to the background. Materials may have low, medium or high suction.

Low suction

Backgrounds like gloss or silk painted surfaces, metal, pre-cast concrete and ceramic tiles are not very **porous**. The fresh plaster is likely to sound hollow when you tap it or even fall off the wall because it will set and cure on top of the surface as there is not enough suction for it to stick. If so, it will need to be treated with a bonding coat or given a key.

Medium suction

Backgrounds like **engineering bricks**, **common bricks**, medium-density blocks, plasterboard and MultiBoard are slightly porous. The plaster will firm up by setting naturally or by evaporation in warmer areas so the background may still need a bonding coat.

High suction

Backgrounds like some types of brick, aircrete blocks and old plaster are porous, so they will suck moisture from the fresh plaster and dry it out too fast. This makes the plaster too difficult to work with as it may be dry before you can make it smooth, or it will crack after you have finished plastering the wall. The suction will need to be controlled with a primer.

KEY TERMS

Suction

– the porosity or ability to absorb water from an applied material.

Adhesion

– the 'sticking' of a material to the background.

Porous (material)

– something that contains tiny holes that allow water to enter or pass through it.

Engineering bricks

– hard dense bricks of regular size used for carrying heavy loads (e.g. in bridge buildings, heavy foundations, etc.).

Common bricks

– bricks of medium quality used for ordinary walling work where no special face finish is required.

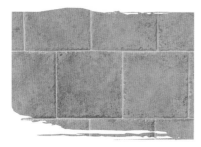

Figure 4.30 A low suction background

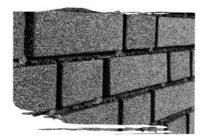

Figure 4.31 A medium suction background

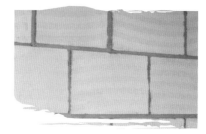

Figure 4.32 A high suction background

Experienced plasterers can tell at once whether a wall is too dry and decide whether additional treatments are needed before the plaster is applied. However, you can do a simple test to find out whether a wall has the right amount of suction. Apply a small patch of plaster to the background, leave it for a few minutes and then test it with your fingers. If it is dry, it is likely to have high suction. If it has not dried out at all, it probably has low suction.

APPLYING PLASTER MATERIALS TO INTERNAL BACKGROUNDS

Preparing background surfaces

Background preparation is one of the most important factors to take into account before plastering begins.

Before applying the plaster, you must take the time to properly prepare the background. If you do not do this, the plaster will not be smooth and flat, and it may even fall off the wall.

First, the background surface should be checked to make sure that it is **plumb** and flat. If the wall is made of irregular blocks, any projecting blocks must be chiselled off. If there are hollows or deep joints in the background surface, you will need to apply a **dubbing out** coat to provide a level flat surface.

The face of the wall should be free from loose dirt and dust, and it must be well dampened to reduce the absorption of moisture from the mortar.

Methods of cleaning down background surfaces

Oil, grease, films, salt, dirt, dust and other loose material can interfere with bonding, so a surface must be cleaned off. Methods depend on how dirty the background surface is, where it is and what it is. They include:

* **brushing down** using a brush or broom

* **scraping** old plaster and dirt off the background surface with a scraping tool

* blowing with (oil-free) **compressed air** but you must ensure that the resulting dust is collected

* **wetting** with a damp cloth or sponge

* **vacuum cleaning** to suck off the loose dust

* **wet sand-blasting** very dirty surfaces

* **water jetting** very dirty external surfaces.

Do not use solvents to remove films formed by curing compounds.

Figure 4.33 Scraping off a surface

Figure 4.34 Water jetting

KEY TERMS

Mechanical key

– grooves or openings made on the background surface of an undercoat plaster through which the top layer of plaster will bond.

Chemical key

– a bonding agent like PVA or a branded primer that is applied to the background surface to help the plaster bond.

Figure 4.35 Raked joint

Controlling suction

To ensure the plaster sticks and is workable, you must choose the right method of controlling suction before the plaster is applied. Adding a background treatment may seem to add extra time to the job, but it will result in a better finish so will save time in the long run.

There are two main ways of controlling suction: providing a **mechanical key** or applying a **chemical key** using bonding agents.

Forming a key to background surfaces

A key provides a rough surface for the plaster to stick to on a low-suction background. This is usually created by raking out brick or blockwork mortar joints to a depth of about 10 mm. Other ways of creating a mechanical key include:

* applying a spatterdash coat (see below)

* cutting zigzags with a craft (Stanley) knife

* hacking the surface with a skutch hammer

* rubbing the surface with a devil float (a float with small nails at one end)

* rubbing the surface with sandpaper or a wire brush

* applying a background treatment containing grit.

Figure 4.36 A spatterdash coat provides a mechanical key

DID YOU KNOW?

If there is a difference in suction between the brickwork and the plaster, you may notice grinning in the wall. This is when mortar joints are clearly visible through the plaster. It is mainly a cosmetic problem – although it looks unsightly, further cracking is rare. You can prevent grinning by applying an undercoat or spatterdash coat before further plaster coats.

Using bonding agents on background surfaces

PVA, EVA and SBR

These are all latex-based adhesives (rubbery glues) and are available in a variety of brands and qualities:

● **PVA** (polyvinyl acetate) is one of the most popular bonding agents because it is fairly cheap and is useful for lots of different purposes.

● **EVA** (ethylene vinyl acetate) is similar to PVA but is for external use. It does not contain chlorine so is seen as more environmentally friendly.

● **SBR** (styrene butadiene rubber) is particularly effective in damp or humid areas as it is water-resistant. It may be used on its own or mixed with cement or water.

To apply the bonding agent, follow the instructions on the packaging. You will need to dilute it with water before painting it on the wall and it may be necessary to add sand to provide a rougher texture. Several coats might be needed before the surface is sealed. You will have to wait at least 12 hours before the wall is ready to plaster but you should also ensure it is still tacky when the plaster is applied.

Figure 4.37 PVA

Figure 4.38 SBR

Water

The simplest way to control suction to a low suction background is to spray or paint water over the surface until it runs down the wall, showing that no more water can be absorbed. It is not a true bonding agent but it will increase suction enough for plaster to be applied. This is only a short-term solution and it is better to use a chemical bonding agent.

Primers and stabilisers

These are specially formulated to form a bond for plaster. They are usually water-based polymers containing fine sand or **aggregate**, which gives them a gritty texture that the plaster attaches to. Various brands are available, and they are applied with a brush or a roller or can be sprayed. They often dry more quickly than PVA but you should plaster when it is still tacky.

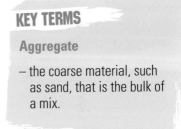

KEY TERMS

Aggregate

– the coarse material, such as sand, that is the bulk of a mix.

Applying spatterdash or stipple coat on background surfaces

Spatterdash

Spatterdash is a type of mechanical key that provides a surface for the plaster to stick to. It is a wet mix (slurry) made of:

● cement and sand mixed to a proportion of 1:1.5 or 1:3, or

● equal parts sand, cement and PVA or SBR

with added water. You can buy it as a dry mixture and add the water when you are ready to use it. The spatterdash is traditionally thrown, or spattered, against the surface, to a thickness of 3–5 mm, using a dashing trowel, a block brush or spraying it on with a machine.

The spatterdash should completely cover the surface and form a rough layer that is then allowed to harden.

Figure 4.39 A primer

Figure 4.40 Spatterdash

Stipple coat

A stipple coat slurry is similar to a spatterdash coat but is mixed from one part of cement and one and a half parts sharp sand before adding water and a bonding agent, such as SBR.

Instead of being thrown at the surface, the mixture is pushed into the surface with a coarse brush and then dabbed with a refilled brush. The resulting coarse finish should be protected from rapid drying out for a day and then left to harden for another one to two days.

Scrim tape

Plasterboard joints should be covered using **scrim** tape – see Table 5.9 on page 166 for more information.

Components used in internal plasterwork

There are different types of metal and uPVC components available to help you obtain the specified finish.

EML

EML (expanded metal lathing) is a sheet of mesh steel to which plaster is applied. It is used internally or externally to:

* provide a key on a variety of backgrounds for materials such as plaster, suspending ceilings and timber-framed buildings

* cover irregular or bumpy surfaces

* cover surfaces where two different materials meet (such as when plasterboard is next to the wood of a new door frame)

* reinforce corners

* cover cracks in weak backgrounds

* cover plastic cable conduits

* make curved and free-form structures

* provide a carrier for fireproof finishes on structural steelwork.

Figure 4.41 Stipple coat

KEY TERMS

Scrim

– self-adhesive tape or jute hessian cloth used to prevent cracking when plastering onto plasterboard.

EML (expanded metal lathing)

– metal reinforcement made out of sheet metal to form a mesh, and fixed over concrete, timber or friable backgrounds to provide a key.

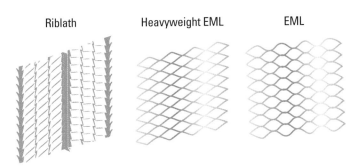

Figure 4.42 Types of EML for internal use (*Source* Expamet)

The flexibility and weatherproof properties of EML give it an advantage over plasterboard in projects where this is important.

A diamond pattern mesh is the most common option. But you can also use herringbone pattern mesh (often known as riblath) or Hy-Rib, which is stiffened by V-profiled ribs between the mesh areas, and which is suitable where long spans between supports are required, such as on ceilings.

EML is made from either galvanised or stainless steel and is supplied in sheets of 2,500 mm × 700 mm, although other weights and mesh sizes are available. It can be cut into the size you require.

Because wood and plaster shrink as they dry, narrow strips of EML may be placed at key positions to minimise cracking. For example, they can be attached over window and door openings, or on ceilings under flush timber beams.

PRACTICAL TIP

When cutting EML, use tin snips or a small hacksaw and ensure you wear protective gloves and goggles. One of the most common injuries to plasterers is catching their hands on the sharp edges of the metal when they cut EML.

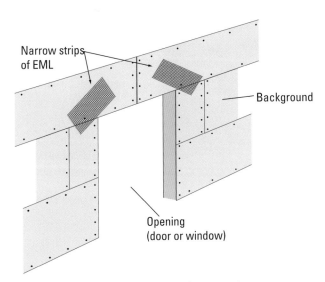

Figure 4.43 Using EML to reinforce crack-prone areas

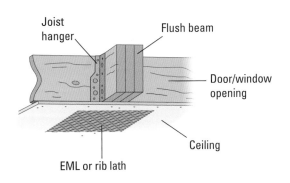

Figure 4.44 Using EML to reinforce ceilings

Beads and trims

These days, you are mostly likely to use **beading** or trims to get a sharp corner, for example where the wall meets the ceiling, on both plaster and plasterboard walls. You will normally attach the beading before plastering the wall, and then plaster over it.

KEY TERMS

Beading

– steel or plastic strips applied to create sharp edges, angles and openings.

Beading has a number of benefits:

* It strengthens corners and edges, so that the plaster is less likely to be chipped.

* Ready-made edges mean that features like **arrises**, **stops** and **movement joints** don't need to be made by hand.

* It is an easier way of providing decorative features rather than trying to mould them yourself out of plaster.

A bead is hollow and flanked by two bands of perforated or expanded metal lath. Galvanised steel or uPVC are used for internal work but stainless steel beading is used where there is a high moisture content. Beading strips come in different lengths (usually 2,400 mm or 3,000 mm) and thicknesses, and are cut to size with tin snips or a hacksaw or nailed together. They can then be bedded into the **scratch coat**.

Figure 4.45 Beading being bedded around a window opening

Table 4.5 describes different types of beading you may need to use for different purposes.

Type of bead	Use
Angle bead (Fig 4.46)	To form external angles and prevent chipping or cracking. These come in different sizes and angles.
Arch bead/former (Fig 4.47)	To form arches and curves around corners or when providing decorative effects. They are normally made from extremely flexible uPVC or galvanised steel.
Architrave/shadow line bead (Fig 4.48)	To form a clean division between different wall finishes, for example around door frames.
Movement bead (Fig 4.49)	To allow for movement between adjoining surface finishes or sections that might move. They usually allow + or – 3 mm of movement. They can also be used where changes in the render colour are specified.
Stop bead (Fig 4.50)	To provide neat edges to two coat plaster or render work at openings or abutments onto other wall or ceiling surfaces. Render (external) versions help with water run-off. You can choose a 9.5 mm or 12.5 mm thickness according to the background you are working to.
Thin coat/skim bead (Fig 4.51)	To be used with thinner coats of plaster, for example over plasterboard where only a skim coat is needed.
Plasterboard bead (Fig 4.52)	To provide a neat edge and reinforce joints and corners. These have similar functions to plaster beads but, for example, stop plasterboard from preventing natural building movement. It is fixed with nails or screws then buried in the skim coat.

Table 4.5 Types of beading and their uses

Figure 4.46

Figure 4.47

Figure 4.48

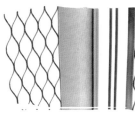

Figure 4.49

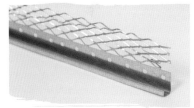

Figure 4.50

Figure 4.51

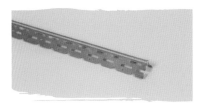

Figure 4.52

Common faults caused by ineffective background surface treatments

If backgrounds are not prepared properly, your plastering will be a waste of time. After you have plastered, there may be problems like movement. This can occur because of structural movement due to the design of the wall or background and may result in cracking. To avoid this, EML may be applied to weak areas, for example where brick and timber meet.

Another common issue is the plaster cracking because a high suction background was not corrected. You should always take the time to treat the suction using suitable methods.

Old plaster can be difficult to remove, especially if it was applied to brickwork using a bonding adhesive.

There are several ways to take off old plaster, depending on how difficult the plaster is to remove.

* Hitting the wall with the head of a claw hammer to loosen the plaster and then using the claw to pull the plaster off. This works best with timber backgrounds.

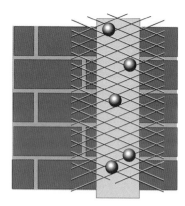

Figure 4.53 EML applied over a joint to allow for movement

* If the plaster is already fairly loose, place the blade of a bolster chisel behind it and hit it with a lump hammer.

* For larger expanses, a spade can be pushed behind the plaster – although this method requires quite a bit of strength.

* For tough plaster and Artex, use an SDS drill with its rotary stop set to the hammer-only setting and with a 40mm tile removal or chisel bit.

Figure 4.54 Using a hammer and bolster chisel to remove old plaster from a wall

PRACTICAL TIP

When removing old plaster with an SDS drill, place the chisel-bit at a 45° angle between the brickwork (or other background surface) and the plaster, pushing the drill away from you along the wall. Using it straight on, at an angle of 90°, will loosen soft bricks in the wall. Some plasterers prefer to start from the bottom and work up to about a metre before removing the loose plaster.

Let smaller loose chunks of plaster fall to the floor but once you've loosened larger pieces, stop the drill and pull them off with your hands. This gives you more control over their removal, and reduces the chance of heavy pieces falling onto your foot or smashing when they hit the ground.

Materials and additives used for internal plasterwork

You can mix your own plaster from sand and cement using a mixer but most plasterers now use bags of pre-mixed plaster powder, often containing useful additives, to which they just need to add water. This is more convenient while on site, at least for smaller jobs, but it is important to know about the materials and additives that go into plaster.

Gypsum plasters

Gypsum is calcium sulphate that has been hydrated. It is the mineral deposit that is left in the rock after the water has evaporated. Gypsum is white, but small impurities colour it grey or pink.

When you add water, it will re-crystallise. This process is known as setting.

Two main types of gypsum plaster are used in plastering:

* **Class A hemi-hydrate gypsum plaster**: Large rocks are crushed to a fine powder and heated to 150°C, producing plaster of Paris. This is mainly used in fibrous work for casting and running moulds, and is also used in the medical industry. It sets very quickly without the aid of a retarder.

* **Class B retarded hemi-hydrate plaster**: This is a class A plaster that has been treated with a retarding agent to slow down the setting time and give it greater workability. These plasters are used as binders in a number of lightweight plasters, and can be used on a variety of plastering work.

Most gypsum-based plasters now use lightweight aggregates like exfoliated **vermiculite** and expanded perlite instead of sand. The aggregates are mixed with Class B gypsum plasters for undercoats and finish coats. They can be used anywhere that sand-based plasters would be used and have better heat resistance and thermal insulation qualities than sand-based plasters, so can be used where condensation may be a problem. They come pre-mixed for different applications, and only need water to be added.

Lightweight plasters are:

* 60 per cent lighter than sand plasters

* 300 per cent more insulating than sand plasters

* fire resistant

* crack resistant

* able to stick to smooth concrete and to glazed or oil-painted surfaces.

In new work, the type of plaster you use depends on:

* the background materials

* the suction of the background

* the required hardness of the finished plaster.

Most plasters are produced by British Gypsum and Knauf, and have recognised brand names that you will soon become familiar with. British Gypsum provides the majority of material for the plastering industry in the UK, but new technology and easier production has meant that other companies are competing to produce materials for plaster that are being used around the country.

Table 4.6 describes some of the plasters that you may use. Even if you do not use British Gypsum plasters, it is useful to know the variety of properties that different plasters can have.

Type of plaster	Name	Purpose
Undercoats	Thistle Bonding Coat	For smooth or low suction backgrounds, e.g. concrete, plasterboard or surfaces pre-treated with bonding agents. You can also use it to fix EML.
	Thistle Hardwall	High impact resistance and quicker drying surface. Suitable for application by hand or mechanical plastering machine to most masonry backgrounds.
	Thistle Tough Coat	High coverage, good impact resistance. Suitable for application by hand or mechanical plastering machine to most masonry backgrounds.
	Thistle Browning	For solid backgrounds of moderate suction with an adequate mechanical key.
	Thistle Dry-Coat	Cement based, for re-plastering after installation of a damp-proof course.
Finish coat plasters	Thistle Board Finish	To skim low–medium suction backgrounds such as plasterboard.
	Thistle Multi-Finish	For use over both undercoats and plasterboard.
	Thistle Uni-Finish	A finish coat plaster that requires no prior preparation with PVA on the majority of backgrounds.
	Thistle Spray Finish	Gypsum finishing plaster for spray or hand application.
One coat plaster	Thistle Universal One Coat	For a variety of backgrounds. Suitable for application by hand or mechanical plastering machine.

Table 4.6 Types of British Gypsum plasters (*Source* The White Book, 2013)

Figure 4.55 Common types of undercoat plaster

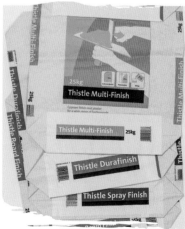

Figure 4.56 Common types of finishing plaster

Non-gypsum plasters

Damp surfaces, or walls that have to be re-plastered after a DPC has been installed, are not suitable backgrounds for gypsum plasters as they will draw salts to the surface of the plaster. Instead, plasters containing well-graded sand and cement should be used. However, these plasters are brittle and dense, and may crack with building movement, so breathable **lime**-based plasters are another alternative. Refurbishment work is likely to use lime plaster, as this is usually the type found in older buildings. These are a 1:1:6 cement:lime:sand plaster mix with a lightweight aggregate used in place of sand.

PRACTICAL TIP

Make sure you use the right sort of sand – its particles must vary in size up to 5 mm and it must be medium sharp. Soft sand, such as the type used in bricklaying mortars, is not suitable.

Types of additives

Cement is a powder that goes through a chemical change or reaction when mixed with water. It becomes like an adhesive paste and then hardens. Its main purpose is to bind bulkier materials like sand and aggregates within the plaster mixture.

Without additives, cement can make plaster shrink and crack as it dries. Too much water in the mix will make the plaster weak. In hot weather the mix can dry too quickly and in cold weather it can dry too slowly or be damaged by frost. Standard mixes are not waterproof and most mixes will develop surface pores as air bubbles escape while the mixture dries, causing weakness and allowing water penetration.

Additives are designed to make working with plaster easier and to produce better results. Additives may have been added to the plaster when it was manufactured or you might need to add them yourself.

Table 4.7 describes some of the additives you are most likely to use.

KEY TERMS

Lime plaster

– a traditional type of plaster consisting of hydrated lime, sand and water and now used mainly when restoring old buildings. It allows walls to breathe, reducing the risk of damp and condensation.

Accelerator

– an additive used to speed up the setting time, for example as a frostproofer in winter.

Plasticiser

– an additive used to make plaster more workable.

Retarder

– an additive used to slow setting and allow more working time.

Additive	Description
Frostproofer/ **Accelerator**	This is a liquid additive formulated to accelerate setting and hardening times of plaster, mortar, concretes, screeds and rendering, so that it is not affected by frost while it is setting. It can be effective in sub-zero temperatures and can also be used in normal temperatures where a rapid set is required. Do not use this additive with lime cement. Add frostproofer to water at a rate of 1.7 litres to every 50 kg of cement.
Plasticiser	This is added to the water to make the plaster smoother and easier to work with. It may be added instead of lime, as it has its advantages but is cheaper and maintains the strength of the mix. Less water needs to be added to the mix, which helps to prevent cracking and shrinkage. It gives the mix additional strength and flexibility. It also often delays the setting time, which makes it suitable for use over large areas. The plasticiser liquid is mixed with water at a ratio of 1:100.
Retarder	This will slow the set time of gypsum plasters, giving you more opportunity to work with the plaster. However, you must take care not to use too much, or it will dry out incorrectly or not set at all. Different types are available but it normally comes in powdered or crystal form and must be dissolved in water before it is added to the plaster.

Additive	Description
Waterproofer	This plasticises the mix by preventing water from penetrating cement plasters without acting as a vapour barrier. It also reduces suction when it is used in cement-based undercoats. Some waterproofers are also retarders. The waterproofer liquid is mixed with water at a ratio of 1:30 and added to the cement/aggregate mix to achieve the required consistency.
Coloured pigments and dyes	These change the colour of the plaster but should have no other effect on it. They usually come in the form of granules which are combined with the dry mix before the water is added. Depending on the depth of colour required, the pigment should be added at a ratio of 2:100 to 10:100. Note that the same quantity needs to be used in different batches to prevent variations in the colour of the plaster.
Expanded perlite	This is a glassy volcanic mineral that contains a small amount of water. When it has been crushed and superheated, this water turns to steam, which helps to form a fine substance that expands its volume up to 20 times. It is fireproof, soundproof and insulating, with a warmer surface than standard gypsum plasters that prevents condensation. It is used in lightweight plasters.
Vermiculite	This is produced in a similar way to expanded perlite, and has similar properties. It has a wide variety of uses but it is used as a lightweight aggregate in plaster applied either by hand or as a spray to improve coverage, ease of handling, adhesion to background surfaces, fire resistance, and resistance to chipping, cracking and shrinkage.
Fibre-reinforced gypsum	This ready-mixed gypsum contains glass fibre, which helps to prevent cracks in the plaster caused by shrinkage.

Table 4.7 Types of plaster additives

Measuring and mixing plaster materials

Procedures for the material mixing area

Before you start gauging and mixing plaster materials, you must make sure that:

* the area is well ventilated, for example with windows and doors open

* the floor is level and the container is large enough for the mix

* you wear appropriate PPE, such as a dust mask to avoid breathing in dust

* there is nothing that can contaminate the mix, for example nearby work that is creating dust

* your mixing equipment is clean

* there is enough space for you to mix the quantity you need.

Gauging plaster materials

To make plaster of the correct consistency, you need to use its constituent materials in the correct proportions. Materials are measured by weight or volume, using the same method for different materials and batches. You can use any suitable container to **gauge** the amount you need, such as a bucket, bag or wheelbarrow, or you can use a **gauge box**. Pour in the materials, level them off and remove the gauge box to leave the correct amounts behind.

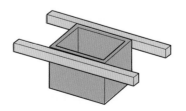

Figure 4.57 A gauge box

Ways of mixing plaster materials

It is much quicker to mix plaster materials with the appropriate tools rather than by hand. However, these should be treated like any other power tool and only used after training and familiarising yourself with the particular make or brand.

Table 4.8 outlines the main tools and machinery you will use to mix plaster materials, and how to look after them.

Tool / machinery	Description	Cleaning and care
Cement mixer (Fig 4.58)	A self-contained mixer is useful for larger jobs. Portable cement mixers have a capacity of 60 to 150 litres. They are powered by electricity (110V or 230V) or petrol.	• Clean mixer by running it with water or water and brick or gravel inside. • Ensure any water is drained after use. • Ensure mixer is switched off after use.
Mortar mill/pan mixer (Fig 4.59)	This grinds and mixes powders, suspensions and pastes and is regarded as more efficient than a cement mixer, especially for the mixing of lime plaster and crushing materials to a lump-free mortar mix. Available in a variety of sizes.	• Always ensure that only materials specified by the manufacturer are placed in the mill. • Clean the mill thoroughly after use.
Whisk/paddle mixer (Fig 4.60)	This is a power tool with a long stem that has a whisk (paddle) attachment at the end, for mixing viscous materials like mortar, adhesive and plaster. It will usually have a two-speed motor for mixing both thin and thick materials, as well as variable speed control. Different types of attachment are available – although they are usually made of rigid metal, new flexible plastic designs are becoming available.	• Clean the whisk by switching it on in a bucket of clean water. • Do not let the mortar on the attachment dry or it will be difficult to remove.

Table 4.8 Powered tools and machinery for mixing plaster materials

Figure 4.58

Figure 4.59

Figure 4.60

Under- and over-mixing plaster materials

As soon as the powder and water meet, a chemical reaction begins that starts the setting process. If mixed correctly, the plaster should last for around 15 minutes, or 20 to 30 minutes in colder conditions, before starting to set.

The longer you mix the plaster, the stronger it will be. However, after mixing it for 10 minutes or so (depending on the mix, mixer, temperature and other factors) it will start to become weaker again. Finding the ideal time you need to take when mixing comes with experience and it will not be the same for every batch.

Under-mixing will result in a plaster that has a lumpy or watery consistency, caused by the powder and water separating. It will not set properly and will be difficult to apply to a wall or ceiling.

Over-mixing – that is, mixing for too long or at too fast a speed setting – will create heat that makes the plaster set more quickly. You might find that you can't spread it on a wall.

Mixing too large a batch could have a similar result – it could start to set before you are ready to pour it.

Methods of applying plaster to backgrounds

Internal plastering may consist of one, two or three coats, depending on the background surface and the type of work specified. The plaster is normally applied in two coats – the first coat (or undercoat) is about 11 mm thick, with a finish coat (or skimming coat) over it that is about 2 mm thick – but the background or function of the plaster may require one or three coats to be applied.

One coat work

All-in-one plasters are sometimes used, and are applied thickly, often using a spray applicator, without a second, finish coat. It may be suitable in old buildings where a completely smooth wall is not required and for patching small areas. However, for standard brickwork walls, most professional plasterers feel that this type of plaster and approach does not give an adequate finish.

You are most likely to use a single coat to provide a thin setting, skim or **finish coat** to plasterboard. It is usually 2 or 3 mm deep and is usually applied in stages. Finish coat plasters are finer than undercoat plasters and can create a very smooth surface – if they are applied with skill. Applying a skim coat is described under 'Two coat work' below, and the procedure is the same as skimming onto a floating coat.

PRACTICAL TIP

One coat work is not the easy option – you need to get the surface both flat and smooth in a single stage rather than two.

Two coat work

This is also known as float and set and is most plasterers' preferred method of internal plastering a solid background like brick or blockwork. It involves applying an undercoat or **floating coat** for a flat surface, which is keyed before a finish (skim or setting coat) is applied to give a smooth surface, free of **gauls** (snots).

DID YOU KNOW?

The process is very similar for both walls and ceilings. The main difference is that you will normally plaster a ceiling at height so you may need to use a scaffold or other access equipment.

PRACTICAL TIP

As we saw earlier, really rough surfaces also need a dubbing out coat to fill up the holes before full plastering begins.

KEY TERMS

Finish coat

– finish plaster applied 2–3 mm thick to provide a smooth finish ready for decoration.

Floating coat

– undercoat plaster, commonly lightweight, applied 8–11 mm thick to a background to make it straight and plumb prior to setting coat being applied.

Gauls

– spatters or blemishes in the setting coat that are not flush to the wall, which must be removed or smoothed to form a flat surface. Also known as snots.

Apply the floating coat to flat surfaces up to a thickness of 5 mm or by hand to 13 mm.

Unlike one-coat plastering, you will be applying your second (finish) coat to plaster (a floated background) rather than plasterboard. This means that, before applying the finish coat:

* you must make sure that the floating coat is not yet completely dry

* you must key the floating coat, preferably with a devil float

* you may need to damp down the background with water if suction is too low

* you may need to scrape down the background, especially internal and external corners, which might need cutting back a few millimetres from the angle, using the back of a trowel.

See below for methods of applying the floating coat – plumb, dot and screed, and broad screed.

Applying the finish coat

Apply the finish coat (also called skim coat, skimming coat or setting coat) with a finishing trowel, starting from the top left side if you are right-handed, and from the right if you are left-handed.

When you apply the plaster, ensure you overlap each trowelful, so that you do not leave any gaps, then go over it to leave a coat about 1 mm thick initially. Then you trowel up by wetting a clean trowel and firmly and methodically smooth the wall, holding the trowel at about 45°. The final coat should be no thicker than 5 mm.

> **DID YOU KNOW?**
>
> A Speedskim is a new tool for ruling off the skim coat. It is like a feather edge rule but has a flexible blade that enables a good finish in a shorter time.

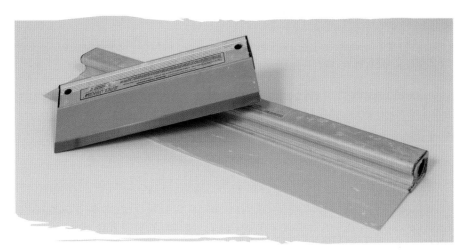

Figure 4.61 A Speedskim

> **PRACTICAL TIP**
>
> Don't worry about the quality of your finish when applying the floating coat – but you are trying to get the flattest surface possible, free of bumps, hollows and curves. If you don't get it right at this stage, your finish coat will not look smooth.

> **PRACTICAL TIP**
>
> When you have a grainy surface like sand or cement, or plasterboard that needs flattening, you can apply an additional thick layer using the float to cover up any irregularities. You can also rule it off with a feather edge rule.

> **PRACTICAL TIP**
>
> A common plaster for two-coat plastering is Thistle Multi-Finish. This is a gypsum finish plaster that can be used on a range of backgrounds. It is popular because it is easy to prepare and apply, and provides a smooth, attractive surface to internal walls and ceilings.

PRACTICAL TIP

Try to avoid working to wet angles by completing the opposite walls first. When these two walls have set the other two opposite walls can be completed.

You should aim to leave a matt finish, as a shiny finish will not provide enough of a key for the subsequent paint. You can do this by brushing a small amount of clean water on the wall in front of your trowel as you finish the skim coat.

When you have finished, clean out all the excess plaster from internal and external angles, the ceiling line, electrical points and the skirting line.

You will often see that plaster is not applied below the skirting line. This is to enable the skirting to be fixed flush to the wall. Mark off a line about 25 mm from the floor and remove the plaster below this line.

Figure 4.62 Applying a skim coat

Figure 4.63 Applying two coat plaster

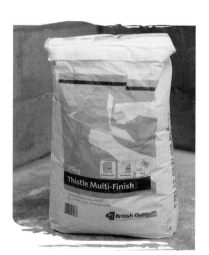

Figure 4.64 Two coat plaster

Three coat work

Sometimes the background requires an extra coat of plaster material. This might be because:

* it is uneven

* it contains metal lathing such as EML, beads and trims

* it is made of several different materials

* the area is severely damaged, for example around walls and windows in older properties

* the plaster needs to be thicker, as specified by the architect or because you need to match an existing thickness.

Three coats consist of two undercoats and a finishing coat, in a process called render, float and set:

1. Scratch coat or render undercoat to provide a solid background and adequate suction. It is keyed and left to dry for 24 hours.

2. Floating coat.

3. Finish coat.

It should be applied as layers of up to 15 mm, which are left to dry before the next layer is applied. Once the scratch coat has set (usually after a few days), you can add the floating coat.

You can apply the floating coat using the plumb and dot method (also known as the plumb, dot and screed method) or the broad (or box) screed method. Plumb and dot is more complicated but is thought to produce a better result than the quicker broad screed method.

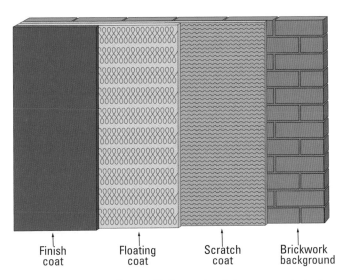

Figure 4.65 Example of three coat plastering

Finish coat | Floating coat | Scratch coat | Brickwork background

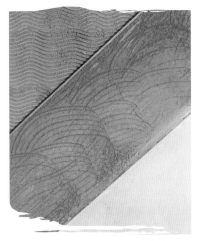

Figure 4.66 Three layers of plaster

Using the plumb, dot and screed method

The plumb, **dot** and **screed** method produces the most accurate result but it is fairly time-consuming.

Before you start, check whether there is a door or opening in the wall. If so, you will need to set your dots to the same level to ensure the wall doesn't go too far back or forward.

Set a dot approximately 150 mm from a bottom corner of the wall, bedding it in with a little plaster. Then set a dot 150 mm from the other bottom corner. It is worth setting a line between them to check whether they are level. If not, adjust the dots.

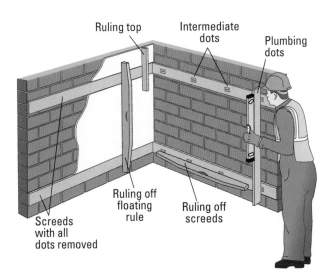

Figure 4.67 Using the plumb, dot and screed method

Ruling top | Intermediate dots | Plumbing dots | Ruling off floating rule | Ruling off screeds | Screeds with all dots removed

KEY TERMS

Dots

– small pieces of material such as plywood applied to the background to help produce a flat floated surface. They form guidelines for the floating rule when applying screeds.

Screeds

– narrow bands of plaster material which are built up between plumbed or levelled dots, to act as guides for the floating rule when creating plumb or level surfaces on walls and ceilings.

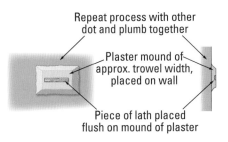

Repeat process with other dot and plumb together

Plaster mound of approx. trowel width, placed on wall

Piece of lath placed flush on mound of plaster

Figure 4.68 A dot

121

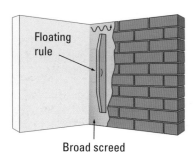

Floating rule

Broad screed

Figure 4.69 Using the broad screed method

Once they have set, apply dots in the top corners and check them with a level and plumb with the bottom dots.

Now set more dots in a row above the bottom dots, using a string line to ensure they are straight. These should be about 1,500 mm apart but ensure your straight edge can reach from one dot to the next. Repeat for the dots below the top row. You should now have four rows of dots.

After they have set, form the screeds by applying a band of plaster vertically between two of the dots, making sure it sticks out further than the dots. With your rule, push the plaster back so that it is flush with the dots and use your trowel to cut off the surplus plaster on each side of the rule. Carefully slide your rule off the screed and continue until you have applied screed between all the dots.

Once the screed has set, you need to fill in the bays (the sections between the screeds). Damp the screeds down to control the suction then float one bay at a time, using your rule as before and cutting off the excess plaster. Fill in any hollows or gaps and rule it off again before going on to the next bay.

Keep checking your wall to make sure it is flat and plumb.

When you have finished every section and it has set, key the wall using a devil float.

Floating to wet screed using the broad screed method

A problem with the plumb, dot and screed method is that you have to wait for the plaster to set, which can cause problems on site when there is a tight schedule. This means you often need to apply your floating coat quickly. This method, of floating and ruling it off, is preferred by many plasterers on site for this reason, but the finish will not be as good a quality.

1. Apply the plaster to a corner section of wall and pass the rule across it until the plaster is consistently around 11 mm thick.

2. Pass the rule over the plaster using an up-and-down movement until it is flat and smooth. If you have missed any areas, fill them in and re-rule them.

3. Trim off any excess plaster from ceilings, angles and wall lines using a trowel and clean these areas with a brush.

4. Repeat for the opposite corner of the wall.

5. Apply the floating coat from right to left if you are right-handed, or left to right if you are left-handed. Rule off the wet screeds using a straight edge. Keep cleaning off the rule to make sure your edge remains sharp.

6. When you have ruled in the wall flat, check it is flat and plumb with a level and correct any irregularities.

7. Key by rubbing up with devil float and clean all angles with a trowel.

Ruling in the top screed

The wet screed

Filling in the bottom screed

Filling out the middle section of the wall

Checking the wall with straight edge

Figure 4.70 Applying a floating coat using the broad screed method

Forming internal and external angles

If you are plastering a room, you will need to know how to apply plaster to its corners (internal angles) and around any parts other wall that stick out, such as a chimney breast or inset window frames (external or hard angles).

External angle

Internal angle

Figure 4.71 Internal and external angles in a room

These can be:

*   **Planted** using pre-formed beads and trims to act as a ready-made angle

*   **Formed** or shaped by hand, using plaster and tools

*   **Run** using a running rule.

The method you choose will depend on the specification and the particular angles you encounter. You may also find that you develop a personal preference over time. The angles will not always be 90° – they could be less or more – so you need to use the solution that is best suited to the wall.

Planting external angles using beading

Fixing beading is the most common and straightforward method of forming an angle. First, prepare by checking that the wall is plumb and level; cut the bead to the right length using tin snips and mix enough plaster to attach the bead.

Fixing an angle or stop bead

Once everything is ready, apply 30 mm thick plaster dabs either as a continuous line or at 600 mm intervals on both sides of the external angles. Press the bead onto the dabs. Check it is straight using a straight edge and that it is level on both sides. Now apply plaster to both sides of the bead and check again for level before cleaning surplus plaster off the bead with tool brush.

If you have to apply beading to a long stretch of wall, you might have to join two angle beads together by sliding a steel rod or dowel into the angled corner.

PRACTICAL TIP

If you don't bed the bead in at this stage, the adhesion may be reduced because it is difficult to squeeze plaster between the bead or mesh and the background surface.

PRACTICAL TIP

When creating external corners, ensure that all angle beads are straight and line up with surrounding walls or frames.

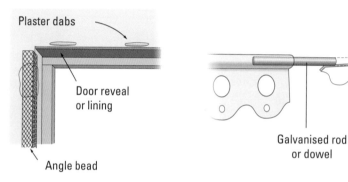

Figure 4.72 Attaching an angle bead to a door reveal or lining

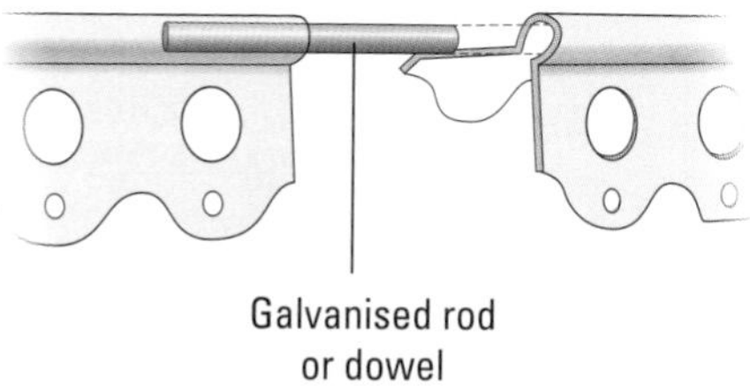

Figure 4.73 Joining two angle beads

Fixing a thin coat bead

Although you can fix a thin coat or skim bead using the method described above, it is more common to attach it using mechanical fixings like steel nails, screws or staples than by applying dabs of finish plaster.

PRACTICAL TIP

Make sure that none of the fixings stick out or you won't get a smooth surface.

Fixing an arch bead

This can be a trickier job as you need to form the curve accurately before fixing it in place. Check that you can make the required shape before fixing it, then screw or nail it on one side before completing it. Again, you can finish it with dabs of plaster. Remember that the curve is unlikely to be flush against the wall, so be careful when plastering over it that the plaster does not push through the holes.

Forming external angles without beading

The easiest way to form an external angle without using beading or trims is to use a timber rule to form first one side of the angle and then the other.

Nail the rule to one of the walls so that it projects beyond the corner to act as a guide for floating the other wall. Float the other wall and, when it has set, remove the rule nail it to the wall you have just floated so that you can treat the second wall in the same way. Remove the rule when the coat has set.

Repeat the process while you complete the finishing coat. When you remove the rule after the finishing coat has set, you might find a raised or rough area behind it, which you will need to remove and smooth down with the back edge of a trowel. However, you might want to leave it while you complete the second wall, as you can use it instead of the timber rule.

Finish the angle using an external angle trowel, Surform or sandpaper to create the shape you require, for example a rounded corner for a pencil round.

You will come across several types of external angle. Here are some of the most common.

A **square arris** can be formed by the **reverse rule method**. Float the wall and then the return angle to a rule then carefully remove the rule. This should leave a floated external angle for rubbing up.

A **pencil round** is formed as a square angle when floating and skimming, and is then rubbed round to the desired shape using a darby or float.

A **bull nose** is best formed with a template but you can form it free hand with a rule and a darby.

KEY TERMS

Reverse rule method

– a way of forming a hard angle or arris with plastering materials without the use of beads.

Figure 4.74 Square arris   Figure 4.75 Pencil round   Figure 4.76 Bull nose

Forming external angles on door or window reveals

Floating to door and window reveals or linings uses similar methods. After checking that the linings are plumb, rule it off with a straight edge held tight to the lining.

When you rub up the wall area around the lining, ensure that the floating coat allows 2–3 mm for the thickness of the skimming plaster. It needs to be the same thickness all the way round so that a consistent amount of the frame can be seen.

One way of doing this is to use a reveal gauge, which can just be a length of timber 50 mm longer than the width of the reveal. When you have formed the corner or fixed the angle bead, place a builder's square with one edge against the horizontal part of the lining and the other edge against the bead or rule. Place the reveal gauge on the square in line with the edge that runs from the frame to the rule or bead.

Floating rule

Figure 4.77 Ruling in a door lining

Now mark the end of the gauge opposite the inner end of the reveal and drive a nail into the gauge, with about 25mm sticking out (or cut a notch into the wood). You can now use the gauge as a horizontal rule, with the nail or notch as a guide for keeping an equal distance from the wall.

You can use this method on larger reveals, such as on attached piers, but if they have a return (change in angle) of 110mm or more, it is easier to apply screeds to the top and bottom using a timber square then use the screeds to rule in the return.

Always clean off the lining when the work is complete.

Forming internal angles

Although you can get uPVC or metal internal angle beads, most plasterers find it easier to create sharp corners using an internal corner trowel. However, this should only be for finishing it off – you need to make sure the angle you form with your floating coat is as straight and true as you can make it using a feather edge or darby.

An out-of-line corner shows up more in a corner. You do not want a wavy line so make sure you are as accurate as possible when applying the floating coat.

A common problem when trying to get a clean angle on corners is 'wet angles' – leaving a trowel line in the wet plaster on the adjoining wall. You will get a better result by plastering opposite walls and letting them set before plastering adjoining walls.

Fixing expanded metal lath (EML) to wall plates and internal work

EML can be attached to timber **wall plates** and lintels, and steelwork. Depending on the background and type of EML used, it can be fixed to the wall using:

* proprietary stainless steel nails

* galvanised 30 or 38mm clout nails

* galvanised 32mm staples

* galvanised wire ties.

Starting from the centre joist, drive in the fixings at 100mm centres then work outwards. Nails should be angled away from the centre joist and bent over.

Try to ensure the sheet is large enough to span the wall plates or joists. Where they meet, overlap the EML by 25mm. If you have to start a new sheet between the joists, the overlap should be 100mm in order to ensure it is strong enough and then you need to wire the sheets together. The EML must be kept tight and fixed in a 'brick bond' pattern with the joints staggered.

All metal lathing requires three coat work in order to ensure there is enough suction and to cover the EML properly. The first (pricking up) coat requires applying the material without touching the metal with the trowel, leaving the plaster interlocked with the mesh of the metal lathing. When the first coat has dried and has been keyed with a scratcher, the floating coat is applied to an average thickness of 9mm and rubbed up with the devil float. It is then skimmed as normal.

> **PRACTICAL TIP**
>
> You fix rib lath in a similar way, ensuring it is at right angles to the joists. As it is stronger than EML, fix it at 600mm centres and allow a larger overlap, of around 50mm. Hy-Rib can be fixed at centres of up to 1,500mm.

> **PRACTICAL TIP**
>
> Plasters suitable for use with internal EML are:
> • Plastalite Expanded Metal Lathing Plaster, which can be used both for the pricking up coat and the floating coat
> • Thistle Hardwall (or similar brands)
> • Thistle Bonding (or similar brands)
> • Thistle Tough Coat (or similar brands)
> • one coat plaster
> • Standard sand/cement plaster.
> Lime plaster is not suitable for use with EML.

Storing unused plaster

Plaster, like food, goes off if it is not used within its shelf life – before the date stamped on its bag. This shelf life should be 3 to 4 months if stored correctly. It will also be unusable if it is stored badly. This is because plaster absorbs water from the air or floor, causing it to set in the bag (air set). Plaster should be stored:

* in the unopened bag it came in

* in an enclosed, well-ventilated room or building

* off the floor, for example on a wooden pallet

* clear of the walls, which may be damp

* in stacks of no more than five

* covered with a sheet or tarpaulin.

Plaster, along with any other bagged materials such as cement, sand, lime, aggregates and pre-mixed renders, should be rotated when new stock comes in. This means putting the newest bags (those with the use-by date that is furthest away) on the bottom of the stack or the back of the store. This should mean that old bags are used before new bags, so that none are wasted.

Figure 4.78 Correctly stored bags of plaster

Protecting the work and disposing of waste

Throughout your plastering job, you should think about other people on site – for example, other trades or the householder if you are working in a domestic setting. You do not want them to damage or spoil your work, and you also do not want anyone to be hurt. In particular:

* Plan your work activities to meet the schedule of works.

* Keep tools and materials tidy and out of the way of anyone who will be in the area.

* Know the setting times of plaster materials you are using.

* Be aware of the atmospheric conditions – that is, the effect of the weather on setting times. For example, plaster may set too quickly on a very hot day, or may set slowly if it is cold and damp.

* If possible, put a barrier around your work area to stop people from entering it.

* Talk to other trades on site so that everyone knows where and when others are working.

* Don't leave mixed plaster materials to set if you are not going to use them.

* Be aware of the dust created by mixing plaster indoors.

Disposing of gypsum waste

In Chapter 3, when we looked at sustainability on pages 82–91 we learned that it is important to ensure that all construction work has the least possible negative effect on the surrounding area. One of the ways of reducing its negative impacts is to keep the site clean and dispose of waste in an environmentally friendly way. This means recycling as much waste as possible.

It is illegal to send plasterboard and gypsum to landfill, mixed in with other waste. This is because gypsum, when mixed with biodegradable waste, can produce hydrogen sulphide gas in landfill. This gas is not only toxic but also smells unpleasant. The Environment Agency recommends other steps to take instead:

* Separate gypsum-based material and plasterboard from other wastes on site so it can be recycled, reused or disposed of properly at landfill.

* Do not deliberately mix gypsum or plasterboard waste with other waste for landfill.

* Find out whether your company's waste management contractor will provide separate skips for the waste or sort it for you.

In any case, all waste, no matter what it is, should be sorted and treated before it is sent to landfill.

CASE STUDY

South
Tyneside *Homes*

South Tyneside Council's
Housing Company

Plastering in the real world

Billy Halliday is a team leader at South Tyneside Homes.

'Plastering is a trade I respect – not everyone can turn their hand to it but you do build up speed and ability with practice.

Some things you can't learn at college. You need constant hands-on experience. Until you get on site, you won't know about things like how much pressure to put behind the trowel, or how to flick it towards you when you're getting plaster off the handboard. There are 101 things you need to know before you even get as far as putting it on the wall. Once you've grasped the basics, you need to learn how to get the right thickness, and get it flat.

It's not until you're at someone's house and you've got to plaster a wall with a window, or deal with bad brickwork that needs dubbing out, that you really start to develop the skills you need for the job. Every job is different – a wall with a hole or one that needs angle beads is going to be different to a straightforward job, and that will affect your pricing up.'

PRACTICAL TASK

1. PLUMB, DOT AND SCREED A SOLID BRICK WALL

OBJECTIVE

To use the plumb, dot and screed method to produce a floating coat that is level, flat and accurate to within 3 mm in a 1.8 m straight edge.

INTRODUCTION

On site, the use of dots and screeds is becoming less popular in favour of the dot and dab method. But this process is an important part of learning how to float and set a wall to the required national standards.

PPE

Ensure you select PPE appropriate to the job and site conditions where you are working. Refer to the PPE section of Chapter 1.

TOOLS AND EQUIPMENT

Plastering trowel	Gauger
Handboard/hawk	Flat brush
Dots (timber laths)	Spot board
Straight edge	Hop-up
Spirit level	Buckets

STEP 1 Soak the timber laths in water and mix the mortar.

STEP 2 Place a small amount of mortar at the top right-hand side of the wall about 150 mm from the ceiling – enough to hold your dot in place. Place a pre-soaked timber lath on the mortar, allowing the lath to stick out about 12 mm from the wall. This is your first dot. Place another lath in a similar position on the left-hand side of the wall.

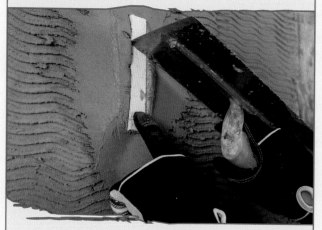

Figure 4.79 Placing the first dot

STEP 3 Now directly underneath each dot place another dot approximately 150 mm from the floor.

STEP 4 Using a straight edge and level, adjust the dots and ensure they are plumb and in line.

Figure 4.80 The first four dots in place

Figure 4.81 A dot flush to the wall

STEP 5 Use a gauge to make sure the dots are set into the plaster at the required thickness and clean off any excess material

STEP 6 Check the dots are sufficiently firm to receive pressure from a trowel. This will take about 20 minutes.

STEP 7 Apply mortar horizontally between the dots, using the dots as a guide. Fill in any hollows to leave two bands of screed ruled out flat and level with each other.

Figure 4.82 Filling in between the dots

Figure 4.83 Ruling off the screed

PRACTICAL TIP

To quicken the process and for training purposes you can add some casting plaster to the material when placing the dots. This will ensure you can rule off without damaging the dots.

STEP 8 Now rub up the screeds using a devil float, filling in any hollows or misses.

Figure 4.84 The screeds in place

STEP 9 Starting from the right-hand side of the wall (or the left if you are left-handed), and using the screeds as a guide, fill in with plaster. Rule off the excess material and check the wall for straight and plumb.

Figure 4.85 Filling in with plaster

STEP 10 When the wall has set, use a devil float to key all areas of the surface to leave a suitable background to receive the finishing plaster.

Figure 4.86 Using a devil float

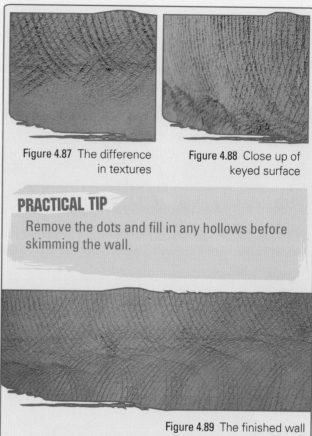

Figure 4.87 The difference in textures

Figure 4.88 Close up of keyed surface

PRACTICAL TIP

Remove the dots and fill in any hollows before skimming the wall.

Figure 4.89 The finished wall

PRACTICAL TASK

2. APPLY A SETTING COAT TO A FLOATED WALL

OBJECTIVE

To complete a wall with finish plaster applied to the correct thickness and trowelled smooth.

INTRODUCTION

This task introduces you to the methods of applying setting coat plasters to a previously floated wall, using various types of finishing plasters.

Figure 4.90 Types of finishing plaster

TOOLS AND EQUIPMENT

Plastering trowel	Hop-up
Handboard/hawk	Buckets
Flat brush	Mixing drill
Spot board	

PPE

Ensure you select PPE appropriate to the job and site conditions where you are working. Refer to the PPE section of Chapter 1.

STEP 1 Check that the floated wall is flat and plumb and has sufficient key.

> **PRACTICAL TIP**
> Sometimes you will find hollows on the floated wall that will require patching before you apply the finishing coat.

STEP 2 Half fill a clean bucket with clean cold water and add the finishing powder slowly.

> **PRACTICAL TIP**
> Allow the water to soak up the powder before you mix with the drill.

STEP 3 Stir thoroughly using a hand or mechanical whisk and ensure all the powder has been mixed in, with no lumps.

Figure 4.91 Mixing the plaster

STEP 4 Clean all the equipment by scrubbing it using a flat brush in a bucket of clean water before you start to plaster the wall.

STEP 5 Start to apply the plaster from the top left-hand side of your wall and work from left to right. Reverse this if you are left-handed.

Figure 4.92 Applying the plaster

STEP 6 Apply an even first coat to the entire wall, approximately 2–3mm thick.

Figure 4.93 The first coat

STEP 7 Allow time for the plaster to dry before applying the laying down coat (second coat of finishing plaster) to an even thickness. Again start at the top left-hand side of the wall.

PRACTICAL TIP

Left-handed plasterers start from the top right-hand side of the wall.

Figure 4.94 Applying the laying down coat

STEP 8 Start the first trowel of the wall (some plasterers call this the flatten trowel). Apply little water at this stage.

Figure 4.95 The first trowel of the wall

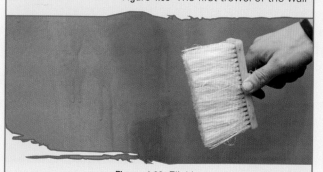

Figure 4.96 Flicking water onto the wall

PRACTICAL TIP

Figure 4.99 shows a plasterer using a prototype Speedskim. These are newly developed tools to help form a smooth surface over a larger area. As they are not yet widely available, you should practise achieving a smooth surface with a trowel before using a Speedskim.

Figure 4.97 Ruling off the wall with a Speedskim

STEP 9 Start the second trowel of the wall. You should have no hollows or trowel marks at this stage.

Figure 4.98 Smoothing off the wall

STEP 10 Now apply the final trowel of the wall. You can add some water with a flat brush to help you smooth the wall so that trowel lines won't show as the plaster starts to set.

Figure 4.99 The final trowel of the wall

STEP 11 Clean all the tools and equipment.

PRACTICAL TASK

3. APPLY TWO COAT WORK USING LIGHTWEIGHT PLASTER

OBJECTIVE

To apply an undercoat of lightweight plaster to a solid brick wall followed by a finishing coat.

INTRODUCTION

The majority of two coat plastering involves using a type of lightweight undercoat plaster and covers a large range of backgrounds. Working to wet screeds is a skill all good plasterers need to learn.

Figure 4.100 Types of undercoat plasters

TOOLS AND EQUIPMENT

Plastering trowel	Buckets
Handboard/hawk	Mixing drill
Flat brush	Straight edge
Spot board	Darby
Hop-up	

PPE

Ensure you select PPE appropriate to the job and site conditions where you are working. Refer to the PPE section of Chapter 1.

STEP 1 Mix the undercoat by adding the powder to half a bucket of water. Ensure the mix is the right consistency before you use it.

STEP 2 Working from the right-hand side of the wall, apply a strip of plaster from wall to ceiling, approximately 11 mm thick. This is your wet screed.

STEP 3 Rule in the material with a straight edge and fill in any hollows.

Figure 4.101 Applying the first wet screed to the wall

Figure 4.102 Ruling off the undercoat

PRACTICAL TIP

Some plasterers tighten in (smooth) the floating coat, filling in the hollows with a darby and smooth off the plaster ready for the devil float.

STEP 4 Carry out the same process on the left-hand side of the wall and then horizontally at the top and bottom of the wall.

STEP 5 Now carefully fill in the middle of the wall with the undercoat plaster and rule in from all the screeds.

STEP 6 Now rub up with a devil float and fill in any hollows. Clean all angles with a trowel.

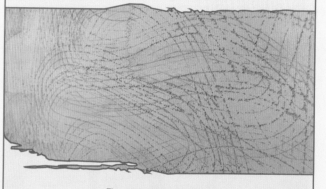

Figure 4.103 Wall keyed with a devil float

STEP 7 Carry out the skimming process by following steps 2–10 in Practical Task 2 above.

PRACTICAL TASK

4. FIX ANGLE BEADS TO BRICK EXTERNAL ANGLE

OBJECTIVE

To fix an angle bead to an external brick wall, using lime mortar, and ensuring that the bead is plumb and level to a ±3 mm deviation.

INTRODUCTION

Angle beads are used to protect the external corner of a wall and form a neat finish to a corner ready for to receive undercoat plaster. The most common beads used on site for two coat work are angle beads and stop beads.

PPE

Ensure you select PPE appropriate to the job and site conditions where you are working. Refer to the PPE section of Chapter 1.

TOOLS AND EQUIPMENT

Plastering trowel	Tin snips
Handboard/hawk	Flat brush
Angle bead	Spot board
Straight edge	Hop-up
Spirit level	Buckets
Gauger	

STEP 1 First check the wall for plumb: use a straight edge and level so you can identify and remove any hollows or high points before you apply the plaster.

PRACTICAL TIP

Dub out any hollows with a little plaster.

STEP 2 Measure the wall and cut the bead using tin snips, remembering wear protective gloves. Check that the bead fits before you start to fix it to the wall.

PRACTICAL TIP

Always store angle beads carefully away from walkways and site traffic. Beads can easily be damaged and if they are bent they will be difficult to use.

STEP 3 Apply plaster dabs along the external angles.

Figure 4.104 Applying plaster dabs to fix beading

STEP 4 Carefully press the bead onto the plaster dabs. Tap it in with the edge of a trowel or float and gently check for straightness by placing the straight edge or level up to the bead.

137

Figure 4.105 Tapping the bead into place

STEP 5 Now, using a straight edge and a level, check that the bead is level on both sides of the external angle. Adjust it where needed by pressing the bead into the wall or out to meet the straight edge

Figure 4.106 Checking for level

STEP 6 Now apply more material to both sides of the bead, tightening in the plaster and filling any missed areas of the wall.

STEP 7 Check the bead again for level before using a flat brush to clean off the bead and remove any excess plaster.

Figure 4.107 The beading in position

PRACTICAL TASK

5. FIX NARROW STRIP EXPANDED METAL LATH (EML) TO A BACKGROUND

OBJECTIVE

To reinforce the background to receive undercoat plaster by fixing strips of expanded metal lath, with a width no greater than 300 mm, over timber lintels or wall plates.

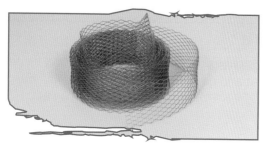

Figure 4.108 A roll of EML

TOOLS AND EQUIPMENT

EML	Tin snips
Screws	Hop-up
Nails	

PPE

Ensure you select PPE appropriate to the job and site conditions where you are working. Refer to the PPE section of Chapter 1.

STEP 1 Measure the area where you need to fix the EML, making sure the mesh will cover the entire timber lintel and any overlaps where the brickwork meets the lintel.

STEP 2 Cut the EML to the required size with tin snips. Cut across the mesh where possible.

PRACTICAL TIP
Remember to wear protective gloves when cutting the EML.

Figure 4.109 Cutting the EML

STEP 3 Place the EML over the area. Starting from the centre of the timber lintel and working outwards, drive in the fixings at 100 mm centres. Angle the nails away from the joist and bend them over with the hammer as you go.

Figure 4.110 Hammering in nails to secure the EML

STEP 4 Make sure the EML is kept tight and flat against the background surface, with all nails are in place and not sticking out of the mesh.

STEP 5 Remove and store safely any small offcuts of EML.

PRACTICAL TIP
Store EML in a safe place, laid flat. When carrying it on site, always wear gloves and fold the edges over so that there are no sharp edges showing.

6. FLOAT AND SKIM TO A WINDOW WALL

OBJECTIVE

To apply a finishing coat to a wall with a window or other opening.

INTRODUCTION

Working to angle beads, you must float a window wall with an undercoat plaster. Ensure the entire area is plumb and straight and window reveal are square to a tolerance of 3mm in a length of 1.8m.

PPE

Ensure you select PPE appropriate to the job and site conditions where you are working. Refer to the PPE section of Chapter 1.

TOOLS AND EQUIPMENT

Plastering trowel	Gauger
Handboard/hawk	Builder's square
Angle beads	Flat brush
Straight edge	Spot board
Float	Hop-up
Darby	Buckets
Spirit level	Mixing drill

STEP 1   Fix the angle beads to the external corners, following the steps in Practical Task 4, or use the same area if it is a window wall. Make sure all the beads are in line and use a straight edge to check accuracy.

STEP 2   Calculate how much plaster is required for the undercoat. You can do this by measuring the total area of the wall and dividing this figure by the coverage stated on the instructions on the plaster bag.

STEP 3   Mix the undercoat by adding the plaster to half a bucket of water. Mix until the plaster is the right consistency.

PRACTICAL TIP

Make sure you use an appropriate undercoat plaster.

STEP 4   Working from the right hand side of the window wall, apply a strip of plaster from wall to ceiling at approximately 11mm thick.

PRACTICAL TIP

Use a longer straight edge for this task, making sure it reaches across the window wall and the area of plaster on each side; this will help when ruling in the floating coat.

STEP 5   Rule in the material with a straight edge, using the angle beads as a guide. Hold the rule tight to the angle beads and keep it straight, filling in any hollows as required.

STEP 6   Carry out the same process on the left-hand side of the wall and then horizontally at the top and bottom of the wall. Again, fill in the hollows and use a darby to smooth off the plaster to a flat finish.

Figure 4.111 Applying undercoat plaster

Figure 4.112 Levelling off the undercoat

STEP 7 Apply undercoat plaster to the window reveals and, using a builder's square against the angle bead and the window, rule in the plaster. Check the margin is the same thickness along the window, using the square or a tape measure.

Figure 4.113 Levelling across the window

STEP 8 Now rub up with a devil coat and fill in any hollows. Clean all angles with a trowel, and clean the external angle beads and window reveals with a small, flat brush.

Figure 4.114 Rubbing up the undercoat with a devil float

PRACTICAL TIP

When rubbing up the wall around the window, you must ensure that the floating coat is scraped back sufficiently to allow for the thickness of the skimming plaster (approximately (2 to 3 mm).

STEP 9 Fill a clean bucket halfway with clean, cold water. Add the finishing plaster powder slowly. Stir it thoroughly using a hand or mechanical whisk until all the powder has been mixed in.

STEP 10 Start to apply the plaster from the top left-hand side of your wall and work from left to right. Apply an even first coat to the entire wall, approximately 2 to 3 mm thick.

Figure 4.115 Applying the finishing coat

STEP 11 Allow time for the plaster to dry and then apply the laying down coat to an even thickness.

PRACTICAL TIP

Left-handed plasterers start from the top right-hand side of the wall.

STEP 12 Apply the first trowel of the wall, applying a little water in front of your trowel.

PRACTICAL TIP

Some plasterers call the first trowel the flatten trowel.

STEP 13 Apply the second trowel of the wall. You should have no hollows or trowel marks at this stage. Make sure you clean around the window and angle beads whenever you need to.

STEP 14 Now apply the final trowel of the wall, applying a little water with a flat brush if you need to. Clean all tools and equipment after use.

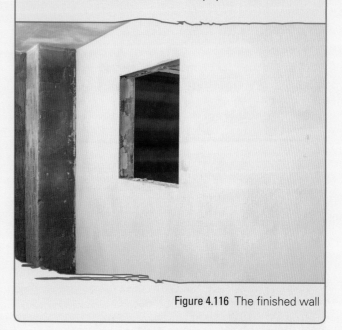

Figure 4.116 The finished wall

PRACTICAL TASK

7. FORM EXTERNAL ANGLES WITHOUT BEADS

OBJECTIVE

To form an external angle using an undercoat plaster and then skim it to a smooth finish, within tolerances of ±3 mm in an 1.8 m length.

INTRODUCTION

Plasterers many need to form an external angle where repair work is necessary or where specified. This task will take you through forming a square arris external corner.

PPE

Ensure you select PPE appropriate to the job and site conditions where you are working. Refer to the PPE section of Chapter 1.

TOOLS AND EQUIPMENT

Plastering trowel	Gauger
Handboard/hawk	Flat brush
Straight edge	Spot board
Timber rules	Hop-up
Float	Buckets
Spirit level	Mixing drill

STEP 1 First check the wall for plumb. Use a straight edge and level so you can identify and remove any hollows or high points before you apply the plaster.

STEP 2 Place the straight edge vertically on one side of the return angle. Make sure it is level, allowing approximately 11 mm clearance to receive the undercoat plaster.

STEP 3 Apply the undercoat plaster to the wall, using the straight edge as a rule. Apply pressure to the trowel as you tighten in the plaster.

PRACTICAL TIP

For the best results, and to free up your hands, ask another student to hold the straight edge while you apply the plaster.

Figure 4.117 Ruling off the first angle

STEP 4 Carefully slide the rule away from its position and place it on the return angle (the adjoining wall on the other side of the angle). Carry out the same procedure on this side, leaving a floated external angle ready to be rubbed up.

Figure 4.118 Applying plaster to the return angle

Figure 4.119 Continuing the angle

STEP 5 Carefully rub up the area with a devil float, always working away from the external angle. Allow to set for 3 hours before starting the skimming process.

Figure 4.120 The completed external angle prior to the skimming coat

STEP 6 Apply the skimming coat, carefully running the trowel away from the external angle on both returns of the wall.

TEST YOURSELF

1. Which regulations require airborne contaminants to be removed from the workplace?

   a. HSE

   b. COSSH

   c. LEV

   d. EML

2. What would you use a feather edge rule for?

   a. Measuring distances

   b. Dubbing out

   c. Levelling off surfaces when floating

   d. Applying PVA

3. What should you use to cut EML?

   a. Tin snips

   b. Scissors

   c. A jig saw

   d. A craft knife

4. What effect does a high suction background have on plaster?

   a. It stops the plaster from setting

   b. It will not bond properly so the plaster sets on top of the surface

   c. It will dry too quickly so the plaster cracks and falls off

   d. It will make the plaster damp

5. Where are you most likely to use SBR?

   a. In very hot areas

   b. In damp or humid areas

   c. On high suction backgrounds

   d. On walls that require lime plaster

6. What is the main feature of an arch bead?

   a. It can be easily moved after plastering

   b. It is pre-moulded in a curve

   c. It is very rigid

   d. It is very flexible

7. What is a floating coat?

   a. An undercoat applied before the finish coat

   b. The coat applied on top of the finish coat

   c. A coat applied using all-in-one plaster

   d. A thin finish coat

8. How can you avoid working to wet angles?

   a. Complete opposite walls first before plastering the adjacent wall

   b. Reduce the background suction

   c. Always plaster adjacent wall at the same time

   d. Use an accelerator in the mix to speed the set

9. Which of these plasters is NOT suitable for use over EML?

   a. Thistle Tough Coat

   b. Manually mixed sand/cement plaster

   c. Lime plaster

   d. One coat plaster

10. What sort of background are you most likely to attach a thin coat bead to?

    a. Fragile brickwork

    b. Plasterboard

    c. Blockwork

    d. Old plaster

Chapter 5

Unit CSA L2Occ54
FIX DRY LINING AND PLASTERBOARD PRODUCTS TO INTERIORS

LEARNING OUTCOMES

LO1/2: Know how to and be able to prepare for fixing dry lining and plasterboard products

LO3/4: Know how to and be able to prepare materials for fixing dry lining plasterboard products

LO5/6: Know how to and be able to fix and finish dry lining and plasterboard products to interiors

INTRODUCTION

The aims of this chapter are to:

* show you how to prepare to use plasterboard products

* help you to select the right plasterboard product for the job

* show you how to fix plasterboard materials, using different methods

* explain how to finish plasterboard and dry lining.

Plasterboard has a variety of applications, from **dry lining** walls and ceilings to forming new internal walls and providing moisture, sound, fire and thermal resistance. It provides a flat surface for plastering or painting. It is used extensively in new build projects, but renovations also make use of plasterboard, for example, when a new stud wall is required.

Other advantages are as follows:

* It provides better insulation than plaster, because of the layered materials it is made from, and because dry lining creates an air gap between the plasterboard and the background surface, which traps heat.

* It is quicker to build a wall from plasterboard than from bricks or blocks.

* You don't need to wait for plaster to dry before applying a finish. For this reason, it is often used as an alternative to two coat plastering work.

All these are economic advantages because the project is finished sooner, and energy bills are lower for the occupier.

KEY TERMS

Plasterboard

– a panel of gypsum plaster that has been pressed between two thick sheets of paper. One side of the plasterboard is used for dry lining and the other side for plastering.

Dry lining

– applying plasterboard to background surfaces like brickwork, timber or metal to provide a flat surface, to secure insulation or to provide the walls with waterproof, fireproof or soundproof properties.

Figure 5.1 The layers of plasterboard

PRACTICAL TIP

Applying plasterboard to a timber frame to form a new wall is not dry lining because it is not attached to an existing background surface.

PRACTICAL TIP

Do not use plasterboard to cover up a damp wall. The problem has not been solved, and the wall could become worse, with the damp showing through the board.

KEY TERMS

Direct bond

– fixing of plasterboards or beads with dabs of plaster or adhesive.

Dry lining

The terms plasterboarding and dry lining are often used interchangeably. Dry lining gets its name because a dry material (plasterboard) is applied to an existing wall, rather than a wet material (plaster). Confusingly, sometimes dry lining refers to the **direct bond** (dot and dab) application method, which uses wet plaster and adhesives rather than mechanical fixings.

KNOW HOW TO PREPARE FOR FIXING DRY LINING AND PLASTERBOARD PRODUCTS

Hazards of working with plasterboard

The main hazards when working with plasterboard are manual handling and the use of hand tools when you cut and fix the board. Check the risk assessment before you begin work to see how these risks have been mitigated (removed or reduced) and to see if there are any more risks related to the specific project.

As always, wear the appropriate PPE for the conditions, such as hard hat, high-vis jacket, safety boots and a dust mask if necessary.

Manual handling

Gypsum-based sheet materials are heavy. A standard sheet of 1,200 mm × 2,400 mm plasterboard that is 12.5 mm thick will weigh around 23.5 kg. This means a pallet of 72 sheets will weigh 1.6 tonnes (*source* wrap.org.uk). You can therefore injure your back or strain your muscles if you do not consider the risks of manual handling.

Plasterboards are delivered either bound in pairs with end tapes or as single sheets.

Plasterboards delivered in pairs are heavy and awkward because of their size, and should be handled by two people.

Where possible, you should use mechanical means to transport boards, such as a forklift truck. But you should only operate this machinery if you have been trained to do so.

Where this is not practical, for example if you are working in someone's home, you will have to carry the boards yourself. Always carry plasterboard on edge, not flat, to avoid putting unsupported pressure on the board's face. If possible, get someone to help you move it.

There are also hazards when installing the plasterboard, especially to ceilings. You may strain your neck or injure your arms by holding boards above your head. You may also need to hold your wrists at an awkward angle when drilling to fix ceiling boards. Mechanical aids, such as ceiling lifts and adjustable props, will reduce the pressure when working with boards above head height. Many workers prefer to use these aids as they generally make the job easier.

Similarly, to avoid straining your back when bending to cut plasterboard, place it on a trestle or other raised surface.

Using tools

You will use a craft knife (Stanley knife) to cut plasterboard to size. You will use a hammer or drill to fix it in place. Although these are fairly simple tools, hazards may arise as a result of them being misused or poorly maintained.

DID YOU KNOW?

Plasterers have one of the highest rates of musculoskeletal disorders (MSD) in the construction industry.

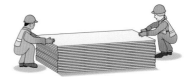

Bend your knees,
not your back
Always lift with
a **straight back**
The sheet only needs to be lifted
a little way off the floor

Figure 5.2 How to carry plasterboard safely

DID YOU KNOW?

For more information on the health risks of installing plasterboard, and recommendations for reducing these risks, see the Health & Safety Executive (HSE) Research Report 812: *An investigation into the use of plasterboard manual handling aids in the GB construction industry and factors helping and hindering the practicability of their application.*

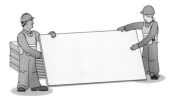

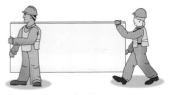

It is your responsibility, and that of your employer, to make sure that any tools you are using are in good condition and safe to be used. The main causes of injury are:

* not using the right tool for the job

* not having been trained to use the tool in a safe way

* deliberately using the wrong tool

* failing to maintain the tool

* not wearing the right PPE.

Tools need to be regularly inspected and some will need routine maintenance to stop them from being potential hazards. Cutting tools, such as craft knives, will need to be regularly sharpened. Wooden handles must be free of cracks and splinters.

If a tool cannot be properly repaired it should be replaced.

When you are carrying out a risk assessment remember that poorly maintained hand tools are more dangerous than tools which have been properly looked after. Tradespeople rarely use poor quality tools, partly because you are more likely to injure yourself with these types of tools.

PPE

The risk assessment will identify how health and safety risks should be reduced to prevent the need for PPE. It may also suggest collective protective measures, such as scaffolds, which protect more than one person. Remember that PPE is always the last resort. However, it is usually still necessary to wear appropriate PPE.

You should always wear eye protection when cutting plasterboard. You are likely to create dust and fragments of the plasterboard may fly off. These fragments are potentially very dangerous, particularly to the eyes.

You should always wear some kind of hand protection too. The thickness of the gloves will depend on the job, as you may need to be able to hold things, which can be difficult if your gloves are too thick. It may seem inconvenient and hinder your work but suitable gloves will protect your hands against cuts, abrasions and most impacts.

Using plaster and cutting plasterboard may require RPE (respiratory protective equipment) such as dust masks.

More information about suitable PPE and the laws concerning it can be found in Chapter 1 on pages 31–33.

Using information sources to fix plasterboard

As we saw in Chapter 2, different types of documentation can help you to plan your work and position the plasterboard or plastering correctly.

Working drawings are either full size or scaled, accurate illustrations of the final finish that is required. From these working drawings you will be able to mark off the plasterboard to the required dimensions to allow you to make the necessary cuts. Accurate positioning following either full-size or scaled drawings is very important. Equally important is where you put the fixings, as you may accidentally drill into cables or pipes if you have not checked their position beforehand.

It is important to follow any supplied documentation to ensure that your work complies with Building Regulations and other requirements of the project. For example, the technical drawings may specify particular types of plasterboard to ensure structural stability, acoustic performance and fire resistance.

You should also refer to the plasterboard manufacturer's technical data sheets to make sure it is the correct material for the area in which it will be applied.

PRACTICAL TIP

Under the Building Regulations, insulated plasterboard must be used internally on all external walls in new buildings. If you are refurbishing a building, any changes must be 'no worse' than the original.

Calculating quantities of the materials you need

Calculating how much plasterboard you need

Look again at Chapter 2 (pages 51–56), which explains how to estimate and calculate quantities of resources. You need to work out the area that needs to be covered with plasterboard – that is, the length multiplied by the height to get a figure in square metres (m^2). If the height is the same throughout, then add up all the lengths and multiply those by the height.

Now you have to work out how many boards you need. To do this, you need to know the measurements of the boards you are going to use. If we take the example of a standard sheet size of 1,200 × 2,400 mm, then that is 2,880,000 mm². Divide this by 1,000,000 to get the number of square metres = 2.8 m².

Then divide the area of the room by the area of the board. This will give you the number of boards required.

So, a room measuring 28 m² will need 10 boards to cover it (28 ÷ 2.8 = 10).

If necessary, round it up to the nearest full number.

Of course, you may need to deduct the area of openings such as windows and doors. However, it is likely that you will need to fix boards above or below a window, so you will need to cut it from a board – therefore, you will still need the maximum number of boards.

Plasterboard is usually cut 15 mm shorter than the height of the room as it should not touch the ground.

PRACTICAL TIP

A 3–5 mm gap will be left between the fixed boards to allow for movement. There is no need to take this into account when calculating quantities, as it will make little difference to the number of boards you need.

Calculating the number of fixings you need

You will also need to calculate the number of fixings you need. Some manufacturers supply tables to help you, based on the size of the board and the spacing of the fixings. Some specifications state you must use screws rather than nails so always check first.

Nails are fixed at 150 mm centres – that is, 150 mm apart. Screws are fixed at 300 mm centres. Divide the length of the plasterboard by the distance apart the fixings will be and add one to secure the end: for example, a 2,400 mm length will need 17 nails or 9 screws. Multiply this by the number of lengths that require fixings.

In practice, nails and screws are usually sold in boxes of several hundred, often by weight, so you just need to make sure you have enough. Remember that boxes of larger fixings contain fewer items.

Calculating the amount of adhesive you need

If you are using pre-mixed adhesive, the packaging or manufacturer's data sheet will state how much area one pack covers. This will depend on the method used to bond the plasterboard. You will normally apply continuous dabs of bonding compound to the perimeter of the board. A 25 kg bag would normally cover about 4.5 m^2, which is only 2 or 3 boards, but you will not be completely coating the entire board with it.

CASE STUDY

South
Tyneside Homes

South Tyneside Council's
Housing Company

Get your calculations right

Gary Kirsop, Head of Property Services at South Tyneside Homes, says:

'It's vital to keep your area clean and tidy, not just for health and safety, but also so you don't risk damaging materials because you've over-calculated. It's critical that you get the right sizes and dimensions – if you get this wrong and end up with too many materials, it is a very expensive mistake. These days the cost of materials in construction is horrendous, and if you've got too many, it's not just the cost of the supplies themselves, but the transportation to move them on, and some companies won't even take them back. Many suppliers will tell you it's too risky because they might already be chipped or damaged.'

KNOW HOW TO PREPARE MATERIALS FOR FIXING DRY LINING PLASTERBOARD PRODUCTS

Using compatible materials

We will see later in this chapter that there are many different types of plasterboard, in different sizes and with different properties. It is important to use only the type of gypsum sheets that are specified. Similarly, you need to ensure you are using appropriate bonding adhesives. Using the wrong type of materials could:

* lead to the boards falling off the background surface

* weaken the wall or ceiling, for example if standard plasterboard is fixed in a damp area

* spoil the finish.

This will result in a poor job that you may have to correct, so it could cost you money and damage your reputation.

Even with the correct plasterboard, you may need to apply a sealant, for example a tanking solution in a shower area before the tiles are laid.

Types and sizes of sheet material

The properties of plasterboard vary according to the additives in the gypsum layer and the weight and strength of the lining paper. You need to make sure you use the correct type of plasterboard to match the specification. As well as standard plasterboard, which complies with fire and thermal standards and regulations, plasterboard with other qualities is available.

Most standard plasterboard has one ivory face, which is suitable for having plaster applied to it, and one brown (reverse) face. The properties of specialist boards are indicated by having a different colour on their face, making it easy to tell what sort of board it is.

Sizes

Plasterboard comes in many different shapes, sizes and thicknesses. It usually comes 1,200mm wide, to suit the standard 600mm stud spacing used in modern housing but sheets can be as narrow as 600mm for use where space is limited. Of course, you can also cut it to size. The thickness depends on its purpose – for example, standard plasterboards are available in a 9.5mm and 12.5mm thickness, but insulating plasterboard can ten times thicker than that.

PRACTICAL TIP

Don't skimp on quality. It won't save you money in the long run as you may have to re-do the job later.

DID YOU KNOW?

Fibre-reinforced gypsum boards (like MultiBoard) are not strictly classed as plasterboard because they do not have a paper facing. However, they are still plaster sheet material that you may be required to fix.

DID YOU KNOW?

Paper facing is generally made from recycled paper.

Brands

There are many different brands of plasterboard and each plasterer has their favourite. The brands listed in Table 5.1 are produced by British Gypsum (Gyproc) but you will also see plasterboard made by companies like Knauf and Lafarge. If a brand is specified by the architect, you must use it. If not, choose the type that best matches the specification.

Type of board	Example brand name	Colour of facing paper	Uses	Available sizes
Standard performance	WallBoard or HandiBoard	Ivory	Suitable for most applications where base fire, structural and acoustic levels are specified. Often enables plaster or decoration to be applied directly	Thickness: 9.5 mm, 12.5 or 15 mm 600 mm × 1,220 mm to 900 × 1,800 mm
Acoustic board	SoundBloc	Pale blue	Complies with Part E of the Building Regulations, or goes further where greater levels of sound insulation are required.	Thickness: 12.5 or 15 mm 1,200 mm × 2,400 mm to 1,200 mm × 3,000 mm
Baseboard	Baseboard	Grey (no facing paper)	This is a cheap plasterboard with no ivory facing or insulation. It can be attached to the wall or ceiling to create a flat surface either for a specialist plasterboard or for direct plastering. It can also be used as a trim at the base of a wall, below the main plasterboard.	Thickness: 9.5 mm
Fire-resistant	FireLine	Pink	Non-combustible glass-reinforced gypsum board, giving increased fire protection.	Thickness: 12.5 mm to 30 mm 900 mm × 1,800 mm to 1,200 mm × 3,000 mm
Glass-fibre reinforced	Multiboard	Varies	Glass-fibre is used between the layers instead of paper. It is suitable for constructing all forms of partition and ceilings, including curves, with fire and impact protection.	Thickness: 6 mm, 10 mm or 12.5 mm 1,200 mm × 2,400 mm to 1,200 mm × 3,000 mm
Impact-resistant	DuraLine	Varies	Board with a high density core for greater strength in public buildings, corridors etc.	Thickness: 15 mm 1,200 mm × 2,400 mm to 1,200 mm × 3,000 mm
Insulating	ThermaLine	Varies	Retains heat and controls vapour.	Thickness: 38 mm to 93 mm 1,200 mm × 2,400 mm
Moisture resistant	Tilebacker	Green Ivory for vapour check boards	In damp areas like wet rooms and showers, and as a base for ceramic tiles.	Thickness: 6 mm or 12.5 mm 1,200 mm × 900 mm to 1,200 mm × 3,000 mm

Table 5.1 Types of plasterboard

Figure 5.3 Standard plasterboard – ivory facing

Figure 5.4 Acoustic plasterboard – blue facing

Figure 5.5 Fire-resistant plasterboard – pink facing

Figure 5.6 Moisture-resistant plasterboard – green facing

It is also worth bearing in mind that:

* some plasterboard combines properties like fire resistance and thermal insulation

* plasterboards usually either have a tapered edge for jointing or skimming, or a square edge for textured finishes

* some types of plasterboard have a special paper facing that helps to control suction.

Preparing backgrounds

Ideally, the background surface should be flat, clean, dry, level and in good condition. This is not always the case so you need to dub out and sand off any bumps before you start to fix the plasterboard. If you are using adhesives, you need to ensure the background has good suction – for example, it is not damp, not in a hot position and is not a painted surface.

Make sure that gas, water and electricity services are identified and pipes and wires are not in the way. They should be pushed into the void or chased (embedded) in the wall where possible.

Setting out to wall and ceiling areas

Setting out means taking measurements and marking the area you need to plasterboard. If you do not spend time making sure this is accurate, your work will not be of a good enough quality. If you are fixing the boards by direct bond, it is important that you accurately set out the vertical centre lines for the application of the bonding compound.

Setting out to a wall
First, you need to identify the wall's high spot – that is, the part that sticks out the most. This is what you must work to so that the plasterboard is flat. Transfer this point to the floor and the ceiling.

Remember that you have to allow for the thickness of both the plasterboard and the bonding compound.

Add this measurement to the marks on the floor and ceiling and transfer it across the room by plumbing a line on the wall. Snap this with a chalk line.

> **PRACTICAL TIP**
>
> The terms 'wallboard' and 'baseboard' are sometimes used instead of 'plasterboard' so make sure you are sure you are using the right type before you start work.

> **DID YOU KNOW?**
>
> All construction work must comply with the UK Building Regulations. This is why the specification normally includes specific instructions on the materials to use. In the case of plasterboard, the relevant parts (approved documents) are B: Fire safety, C: Resistance to contaminants and moisture, E: Resistance to sound and L: Conservation of fuel and power (thermal performance of external walls).

> **PRACTICAL TIP**
>
> Allow 10 mm for the adhesive and add the thickness of the plasterboard e.g. 12.5 mm – giving a total of 22.5 mm. Half a millimetre is hard to measure so round this up to 23 mm.

PRACTICAL TIP

Plan your layout so that vertical and horizontal plasterboard joints are staggered and do not line up.

DID YOU KNOW?

Deviations from straightness must not exceed the values given in BS 8212 (Code of practice for dry lining and partitioning using gypsum plasterboard) and BS 8000 (Part 8: Code of practice for plasterboard – Partitions and dry linings). These are:
- Setting out (from intended position): ±3 mm
- Verticality (plumb): ±5 mm
- Deviation in finished surface (flatness, high spot to low spot): 10 mm.

These are generous tolerances so you should aim to be well within them.

PRACTICAL TIP

You should regularly check that your squares and levels are still accurate. Replace any tools that are damaged.

Mark the width of the plasterboard from an internal corner or opening (e.g. 1,200 mm), and allow 25 mm for the joints between the boards. Then mark the centre of each board and plumb down them using a chalk line – this is where you will put the bonding compound.

If you need to work around an opening like a window, mark the width of the boards from its vertical external angles towards the internal corners of the room on both sides of the window.

To get a neat edge, the tapered edge of the first board should slightly overhang the vertical external angle.

Whether or not there is a window, you will probably need to cut boards to fit. Consider how you will place these cut boards in the gaps.

The practicals on pages 169 and 171 describe in more detail how to set out to a wall and ceiling.

Using datums and transferring levels

The plasterboard must be flat, plumb and square.

One way of checking that it is aligned is to mark datum points using a level. (Table 7.3 on page 201 shows the main types of levels you will come across.)

Setting out levels

In Chapter 2, we looked at using datum points as a reference to creating a level area. On a construction site, this should already have been worked out and marked. If this has not happened, or if you are at a private house, you will need to work out your own datum level, for example the bottom of a door. This is marked and then transferred around the building to wherever it is needed.

You first need to mark the datum points at regular intervals around the room and then join them up using a chalk line.

Instructions for using the different types of levels, which include laser, automatic and water levels, together with illustrations, can be found on pages 208–209 of Chapter 7.

Protecting the plasterboard from damage

Although plasterboard is a robust building material, it is easily damaged by poor handling and storage. Although it is flexible, it will crack or break if it is bent beyond its stress limits. The edges and ends of the boards are susceptible to damage if dropped or struck by a hard object.

You should take certain precautions to ensure plasterboard and gypsum sheet materials remain in good condition while you are not using them.

- It is your responsibility to store the plasterboard out of the way of construction traffic and other trades. You should not expect an electrician or plumber to have to move boards out of their way.

- Plasterboard comes in pairs, with the ivory face turned inwards to protect it. Keep it like this until you need to use it, so that the facing is not damaged. However, you should take it out of any plastic wrappers, as moisture can easily become trapped in contact with the boards.

Figure 5.7 Stacking plasterboard on bearers

- Always lie plasterboard flat in a neat stack – do not lean it against a wall as this could bend it. Plasterboard is also heavy and could injure you if it falls on you.

- Make sure the floor is flat and strong enough to take the weight of the plasterboard. Take special care with suspended floors.

- Standard plasterboard is easily damaged by moisture so ensure it is always stored somewhere dry, preferably indoors. If you have to store it outdoors, keep it off the ground on a level platform that is at least as long as the width of the boards with supports (bearers) not more than 400 mm apart. Cover it all with a tarpaulin to protect it from rain and moisture.

- Stack the boards by type – don't mix up (for example) standard boards and moisture-resistant boards.

- The stack should not be higher than about 900 mm, or there may be too much pressure on the boards at the bottom. It may also topple over.

- Never stack any other materials on top of the boards.

- Don't slide plasterboard – make sure you lift it cleanly.

DID YOU KNOW?

Piling plasterboard too high is likely to damage it. 10 mm-thick board weighs 6.5 kilograms per square metre, so 40 sheets of 3,000 mm × 1,200 mm board weigh just under 1 tonne.

KNOW HOW TO FIX AND FINISH DRY LINING AND PLASTERBOARD PRODUCTS TO INTERIORS

Materials for fixing and finishing

Plasterboard can be attached to the background surface using either adhesives or mechanical fixings like clout nails or drywall screws. After it is in place, you may also need to reinforce and seal any joints, and seal the boards ready for decoration.

Mechanical fixing aids

Table 5.3 describes the different types of screws, nails and plugs you might need.

Fixing	Description
Plugs (Fig 5.8)	Plastic wall plugs are screw fixing devices. They are usually made either from nylon or polythene. They are colour-coded to match screw gauge sizes. Different types are available depending on the type of plasterboard and background. For example, to fix thermal plasterboard to masonry, you can use a plug that is a combination of masonry nail and plastic wall fixing with an expanding tip and countersunk head.
Clout nails (Fig 5.9)	There is a wide variety of different types of nails, but you will use clout nails for fixing plasterboard. They are usually made of non-corrosive galvanised steel, with a short shank and large, flat head to provide a secure fix. Due to the likelihood of nails 'popping' out of the plasterboard, screws are generally seen as a better option.
Screws (Fig 5.10)	As with nails there is a wide variety of different types of screw but for plasterboarding you are most likely to use drywall screws. These have a flat head, so are less likely to get pulled through the plasterboard than screws with wedge-shaped heads. They are also thinner than some other screws, so are less likely to split the wood they are attached to.
Other plasterboard fixings (Fig 5.11)	Other plasterboard fixings are available, for example for fixing insulated plasterboard sheets to brick and block walls. These consist of a galvanised mechanical anchor with a concave head. They are often fire-proof. You might also use long screws which require special plugs, like those in the photo.

Table 5.3 Types of mechanical fixing used in plasterboarding

Figure 5.8 Plugs

Figure 5.9 Clout nails

Figure 5.10 Screws

Figure 5.11 Other plasterboard fixings

Adhesives

Plasterboard adhesive (also called bonding compound) is usually a gypsum-based powder that is mixed with clean water to bond it to brick, block, sand and cement backgrounds. Standard plasterboard adhesive is not waterproof and is not strong enough to be used on ceilings. It has a typical working time of about 100 minutes.

Using bonding compounds for fixing plasterboard products

Fixing plasterboard using adhesives is known as direct bond, or the dot and dab method.

The application of plasterboard has dramatically changed over the last 10 years. Traditional methods of two-coat plastering are gradually being replaced by the system of dabbing plasterboards to walls. The setting time for the wet materials used in two coat work has sometimes delayed the completion of the buildings, so to speed up construction, manufacturers have designed various dry-lining systems that can be installed quickly and covered with plasterboard.

Direct bond has two other main advantages over mechanical bonding:

* It can be bonded directly to bare bricks, instead of having to fix it to wooden battens

* You do not need to make good the screw holes by covering them over.

These advantages speed up the job without compromising on quality.

You need to have a fairly flat wall to use direct bond – if not, consider using mechanical fixings attached to timber battens.

The type of adhesive you use for the job depends on the type of application and the desired completion time. You will normally use powdered pre-mixed drywall adhesive, which is suitable for all types of background and suction, although it is still good practice to paint the wall with PVA or another primer first. Following the instructions on the package, add it slowly to water and mix it in a well-ventilated area with a heavy duty drill. Leave it to stand for a few minutes and give it a stir before using it.

> **DID YOU KNOW?**
>
> Gyproc Dri-Wall Adhesive is a gypsum-based adhesive for plaster and thermal boards.
> * Suitable for high, medium or low suction backgrounds, as long as they are clean and dry.
> * Supplied in 25 kg sacks.
> * Coverage is 25 kg per 6.25 m² board.
> * Shelf life is 3 months.

> **PRACTICAL TIP**
>
> It is better to use mechanical fixings on a ceiling; otherwise you will have to secure the plasterboard in place with a prop while the adhesive dries, and there is still the risk that it will drop off.

> **REED TIP**
>
> A good tradesperson isn't a tradesperson who never makes mistakes; a good tradesperson is a tradesperson who'll make a mistake, learn from it, and get over it.

Using hand and power tools for mixing and fixing

The tables below describe the tools and equipment you are likely to come across when fixing plasterboard.

Drawing and measuring equipment

Measurement or marking tool	Description
Callipers and dividers (Fig 5.12)	These are used to set out measurements.
Combination square (Fig 5.13)	This is a useful tool that can do the job of a mitre square, tri- square and spirit level. It is useful for checking angles.
Erasers (Fig 5.14)	You will need an eraser to rub out any pencil marks that you need to change.
Levels (Fig 5.15)	Levels are for checking the accuracy of work vertically and horizontally. Spirit levels contain a bubble of air in water – when the bubble is central, thecome in various sizes. Laser levels are an increasingly popular as they are accurate and easy to use. It is useful to have both available.
Marking knife (Fig 5.16)	These are considered to be better than a pencil as they are more accurate. They provide a slight cut that is useful when you get to the stage of cutting the plasterboard.
Pencils (Fig 5.17)	These are graded according to hardness or softness of the lead. You could use a carpenter's pencil but a normal pencil is fine as long as you keep it sharpened to provide a clean line.
Rules and measures (Fig 5.18)	Various options are available. Standard steel tape measures, folding rules and metal steel rules are all used. Although imperial measurement (inches) has been replaced by metric measurement (millimetres), it is wise to have rules and measures that show both. Make sure you don't get them mixed up.
Sliding bevel (Fig 5.19)	This is an adjustable tri-square. It is used to mark out angles other than those of 90°. You can set the blade at a particular angle and then lock it into place.
Tri-square (Fig 5.20)	These are used to mark and test angles at 90°.

Table 5.4 Types of drawing and measuring tools used in plasterboarding

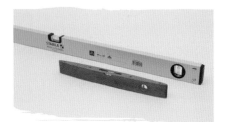

Figure 5.12 Callipers and dividers

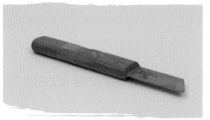

Figure 5.13 Combination square

Figure 5.14 Erasers

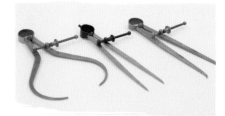

Figure 5.15 Levels

Figure 5.16 Marking knife

Figure 5.17 Pencils

Figure 5.18 Rules and measures

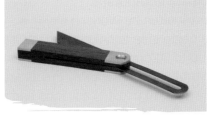

Figure 5.19 Sliding bevel

Figure 5.20 Tri-square

Hand tools

It is useful to keep your usual plastering tools, such as trowels, handy for applying mortar and adhesives, but Table 5.5 shows specialist hand tools for cutting and fixing plasterboard.

Equipment	Description
Circle cutter	This is a hole punch on a long arm, so that you can cut round holes cleanly and accurately.
Drywall/plasterboard hammer	A steel hammer designed to ensure the correct nailing of plasterboard. The hammer end is patterned to dimple the plasterboard to help you position fixings. The hatchet end can be used to score or mark the board, and the notch can remove nails.
Drywall saw (Fig 5.21)	This saw has a sharp point at the end for piercing plasterboard. Some models have blades that lock at different angles, and others have teeth on both sides of the blade. It is usually about 150 mm long, so can be used for making small openings such as for sockets.
Knife (Fig 5.22)	This is most likely to be a retractable Stanley-style utility knife, for scoring plasterboard and cutting thinner boards.
Plasterboard rasp (Fig 5.23)	A flat, stainless steel hand-sized tool for planing and shaping plasterboard. It can be used on edges or in tight spaces and will not clog or collect dust. It may also be called a Surform.
Scoring square	A metal rule with a handle to help score plasterboard accurately with a utility knife. Rather than measuring lines twice and joining them up, you can mark both at the same time.
Taping knife/ applicator (Fig 5.24)	This is a wide steel blade with a handle, used for spreading joint compound over nails, tape, holes and indents. It is usually 100 mm wide but is available in larger or smaller sizes.

Table 5.5 Hand tools for cutting and fixing plasterboard

Figure 5.21 Drywall saw

Figure 5.22 Knife

Figure 5.23 Plasterboard rasp

Figure 5.24 Taping knife/ applicator

Ancillary equipment

You will often be working alone, so you cannot depend on someone else being able to help you reach high areas or fix heavy plasterboard. Trying to lift or hold plasterboard yourself can lead to back injuries. Equipment can take away the strain; however, always make sure you know how to use it in advance and never take unnecessary risks.

Equipment	Description
Board carrier	Several tools have this name, and all are intended to reduce strain when moving boards.
	One is an M-shaped handle that looks a bit like the end of a garden fork. You hold one of the rubber grips and put up to two boards between the other two.
	Another is like a large plastic clip, while you could also use a carrier with a 450 mm handle and a u-channel at the bottom that allows you to slot boards in and carry them like a suitcase.
Board trolley/drywall trolley	These robust steel-frame trolleys are designed for easy handling of sheet and board materials. They are narrow enough to easily go through a doorway and are ideal for areas with uneven or soft flooring.
Foot lifter (Fig 5.25)	This slides under the plasterboard so that you can lever it off the floor to a height of 60 mm or so.
Panel lifter/board lifter	This is a mechanical device to allow one person to lift plasterboard and panels into position onto ceiling and wall studwork. Larger ones can support 75 kg.
	The board can be raised up to 4 m, swivelled through 360° and tilted to 45°. Often they have a solid frame hat can be dismantled for transport.
Panel prop/dead man/panel jack	This telescopic prop or jack holds plasterboard in position, freeing your hands to fix it without having to take its weight yourself. Their adjustable height allows you to use it in different types of room. It can normally support up to 45 kg. Some have a sprung quick release trigger and head that springs the top plate into position after triggering the release button.
Plasterboard fixing tool (BoardMate)	This is a pair of plastic supports that you screw into either end of the plasterboard to hold it in place while you insert permanent fixings. You then unscrew them for use on the next board. It reduces the strain of holding up the board, and can make plasterboard fixing a one person job.
Step-up/hop-up (Fig 5.26)	These are lightweight aluminium platforms for short-duration work on walls and ceilings. They are normally about 80 cm high and about 150 cm long. Larger ones can support up to 150 kg. Models with handrails are safer.
Stilts (Fig 5.27)	These give you some extra height to reach upper walls and ceilings. They are made to fit people of all sizes and have adjustable heights, calf straps and heel plates. Common heights are 450 mm to 760 mm and 610 mm to 1,000 mm.

Table 5.6 Ancillary equipment used when fixing plasterboard

Figure 5.25 Foot lifter

Figure 5.26 Step-up/hop-up

Figure 5.27 Stilts

Portable power tools

Portable power tools can speed up many routine tasks but you must be aware of the risks of using them. According to the Health and Safety Executive, a quarter of all reportable electrical accidents involve portable power equipment.

Portable power tools should only be used by those who are competent, so before using them you should have training and know exactly how to operate the tools. If you do not then you will be putting yourself at risk and possibly others who are working near you.

Figure 5.28 Jig saw

Figure 5.29 Drill

Figure 5.30 Electric screwdriver

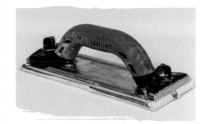
Figure 5.31 Drywall sander

Figure 5.32 Screw or nail gun

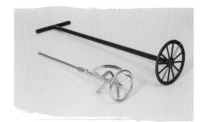
Figure 5.33 Whisk

Portable power tool	Description and use
Jig saw (Fig 5.28)	You would normally cut plasterboard with a knife or a hand-held saw but you might use a jig saw if you have to cut a lot of plasterboard or if a shaped or curved cut is required. The blade cuts on the upward stroke on most machines, but more expensive versions have an action that moves the blade into the material on the upward stroke and away on the downward stroke. This minimises wear and tear on the blade. A number of different blades are available.
Drill (Fig 5.29)	Drills are battery or mains operated. They usually have two different settings and speeds, depending on what you are using the drill for. You can attach many types of bit to it, such as different sized drill bits and screwdriver attachments. Some have a hammer action that strikes into solid backgrounds like concrete. Use the speed that is most appropriate to what you are doing. Slow speeds make a more accurate hole and prevent damage to begin with, before moving up to a faster speed.
Electric screwdriver (Fig 5.30)	Although you can use some drills with a screwdriver attachment, you may prefer to buy a separate electric screwdriver. They are designed to work at lower speeds than drills, which makes them more controllable. You don't usually need to pre-drill the holes for the screws.
Drywall sander (Fig 5.31)	This tool has large diameter discs for sanding drywall joints, painted wall surfaces and ceilings. Hand-held sanders are used for small areas. Each type of sander leaves a more or less smooth surface, so it is important to use the right belt or paper for the job and the material being sanded.
Screw or nail gun (Fig 5.32)	This is used to install dry wall by driving screws or nails into plasterboard. It can also be used with timber or metal. It looks like a normal drill but has a blunt 6 mm tip and may auto-feed screws or nails from a clip.
Whisk (Fig 5.33)	This is a power tool with a long stem that has a whisk (paddle) attachment at the end, for mixing viscous materials like mortar, adhesive and plaster. Different types of whisk attachment are available but you will use a heavy duty whisk attachment for mixing undercoat plaster and dry wall adhesive.

Table 5.7 Types of power tool used in fixing plasterboard

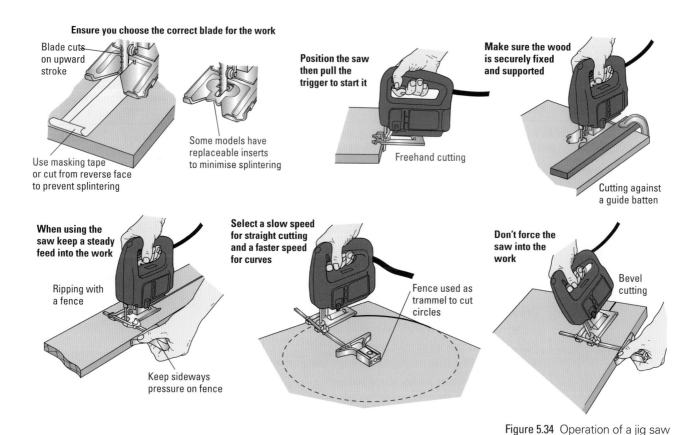

Ensure you choose the correct blade for the work

Blade cuts on upward stroke

Use masking tape or cut from reverse face to prevent splintering

Some models have replaceable inserts to minimise splintering

Position the saw then pull the trigger to start it

Freehand cutting

Make sure the wood is securely fixed and supported

Cutting against a guide batten

When using the saw keep a steady feed into the work

Ripping with a fence

Keep sideways pressure on fence

Select a slow speed for straight cutting and a faster speed for curves

Fence used as trammel to cut circles

Don't force the saw into the work

Bevel cutting

Figure 5.34 Operation of a jig saw

No gloves are worn in these pictures, in order to clearly show how to use hand tools. However, you should wear gloves and other PPE required by your college or employer.

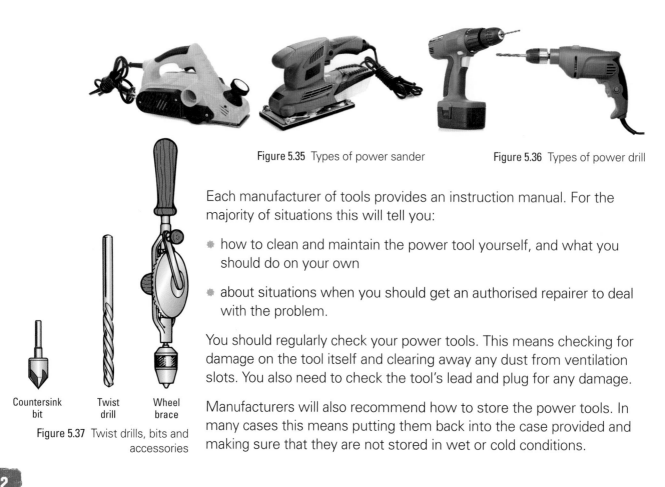

Figure 5.35 Types of power sander

Figure 5.36 Types of power drill

Countersink bit

Twist drill

Wheel brace

Figure 5.37 Twist drills, bits and accessories

Each manufacturer of tools provides an instruction manual. For the majority of situations this will tell you:

* how to clean and maintain the power tool yourself, and what you should do on your own

* about situations when you should get an authorised repairer to deal with the problem.

You should regularly check your power tools. This means checking for damage on the tool itself and clearing away any dust from ventilation slots. You also need to check the tool's lead and plug for any damage.

Manufacturers will also recommend how to store the power tools. In many cases this means putting them back into the case provided and making sure that they are not stored in wet or cold conditions.

Methods of fixing plasterboard

Cutting plasterboard

You should have calculated the amount of plasterboard you need but you will usually need to cut boards to fit the shapes of the room. Plasterboard is normally cut 15mm shorter than the height of the room so that it does not touch the ground, and to allow for skirting boards. However, check the specification to ensure that this is required. You may also need to cut openings into the board for sockets and switches.

Cutting and trimming plasterboard is straightforward but accuracy is essential. Measure the height and width carefully, mark it in pencil and hold a straight edge in place as you score along the line with a knife. Cut only though the face of the paper, not into the plaster inside.

Turn the board over and tap or bend it along the line, using a straight edge, your knee or flat hand, until it snaps. Score along the paper on this side until the piece comes away. Then smooth the cut edge with a rasp or sureform.

If you are cutting both the height and width of the plasterboard, make one cut right across the board before making the other cut.

You can cut out openings either before or after the plasterboard is fixed, whichever is easier. If the socket boxes or light switches are already embedded, you can cut around them after the board is fixed; otherwise you need to measure the position of the holes and cut them out first. Use a pad saw or drywall saw to make the hole and cut it out.

Fixing plasterboard using mechanical fixings

Mechanical fixings are used to secure walls and ceilings. You would not normally drill directly into the wall; the boards are attached to timber battens, joists or a stud wall frame.

Remember to stagger the plasterboard so that you don't have a long line of joints, which could weaken the wall or ceiling. The boards should be placed lengthways to lie along horizontally across joints. Leave a gap of up to 5mm between each board.

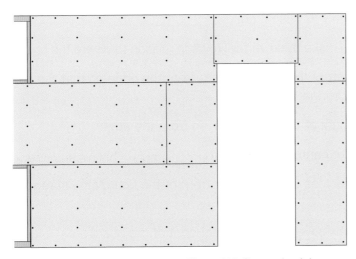

Figure 5.39 Staggering joints across the wall

Figure 5.38 Cutting plasterboard

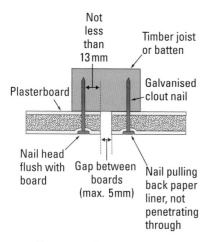

Figure 5.40 Correct positioning of nails

Each board has two bound edges and two unbound edges. Make sure the long bound edge spans the joists at right angles and that the unbound edge is halfway on the furthest joist. This should keep the boards square.

Start fixing from an internal corner and work across to the external angles.

The size of the screws or nails you need depends on the thickness of the plasterboard. They need to attach firmly to the wooden or metal frame behind. Table 5.8 shows the relationship between the thickness of the board and the length of the screw.

Thickness of plasterboard	Drywall screws	Plasterboard clout nails
9.5 mm	32 mm	30 mm
12.5 mm	38 mm	40 mm
15 mm	38 mm	40 mm
19 mm	41 mm	50 mm
12.5 mm double boarded	51 mm	50 mm

Table 5.8 Size of fixings to be used with different thicknesses of plasterboard

Fix screws at 300 mm intervals (centres) and nails at 150 mm intervals, and position them more than 13 mm from the edge of the board.

Do not drive in the head of the nail too much or too little as this could risk puncturing the paper and damaging the plaster inside, so that the board is too loose or falls off. The nail should go in just enough to grip the paper.

The practical task on page 171 shows you how to fix plasterboard using mechanical fixings.

Fixing plasterboard using dot and dab (direct bond)

Mix up the adhesive and put a trowel full onto your hawk board. Pick up a third of a trowel or more and start applying it to the wall around the perimeter of where each board will go. The dabs should be shaped like 'fillets' – about a trowel-length long, and thick (40 to 50 mm) at one edge and thin at the other, like a wedge. This is so the plasterboard can settle when you push it on.

You should make three vertical rows in relation to where the board will go:

* on the left of the board (25 mm from the edge so that the adhesive doesn't bridge the joint)

* on the right of the board (25 mm from the edge)

* down the centre of the board.

Add dabs across the top and bottom of the board area at regular intervals.

Ensure dabs cover at least 20 per cent of the board area, especially when you are going to use heavier boards.

Allow for skirting fixing by applying a continuous line of adhesive where the skirting is going to be. Skirting itself is normally fixed mechanically. You should do the same just below ceiling level, and for services and openings, to ensure they are airtight.

When you are ready to fix the plasterboard, ensure you have packing strips (for example, plasterboard offcuts) under it so it is 15mm off the ground. The gaps will be covered by the skirting.

Press it against the plasterboard and tap it with a straight edge so that it lines up with the chalk lines you made while setting out. Then use a foot lifter to raise the board until it is flush against the ceiling. You will need to use some more packing strips to keep it in place after you remove the foot lifter. Remember to check for level and plumb.

Repeat for the next boards. Their edges should be lightly butted (joined) together. You should have made allowances when setting out and cutting your boards for gaps between joints.

If you have to board an external angle, for example at a window or pier, apply the dabs close to the angle on both sides. Position the cut board edge to the inside, making sure you taper it back so it lines up with the tapered edge. If you don't do this, the jointing will be more difficult.

If you need to nail in plug this is how to do it:

* Wait for the dabs to set.

* Drill two holes per board, about 20mm from the edge of the board and at mid-height. The hole should be 5mm longer than the plug.

* Drive the plugs in so that their heads are slightly below the board without cracking it.

The practical task on page 168 shows you how to fix plasterboard to walls and around windows by direct bond.

Providing secondary fixings

To provide a secondary fixing after you have fixed the board using dot and dab, it is best to use nailable plugs:

* When the dabs have set, insert two plugs per board. Fix them approximately 20mm from the edge of the board and at mid-height.

* Always drill a hole first, making sure it is about 5mm larger than the plug.

* Drive each plug in until the head is slightly below the liner without fracturing it.

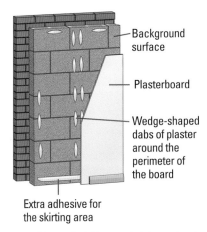

Figure 5.41 Dot and dab method

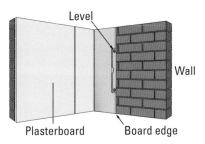

Figure 5.42 Checking board edges for plumb

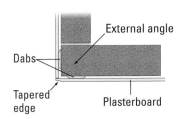

Figure 5.43 Forming an external angle using dot and dab

Joint reinforcement

Table 5.9 describes some of the scrims and tapes you might use with plasterboard. Some types are self-adhesive (sticky-backed) but others are not and need to be held in place with adhesive or mortar before the joint is filled.

You could also use angle beads, which are quicker and easier to use than corner tape because they have a pre-formed corner, so you just need to cut them to length. They are more expensive than tape but you may decide, as most plasterers do, that the extra cost is worthwhile for the better finish.

Type of tape	Description
Corner tape (Fig 5.44)	Used for taping and jointing. Paper joint tape bonded to two corrosion-resistant steel strips for reinforcing internal and external angles in plasterboard.
Glass fibre mesh tape (Fig 5.45)	Strong, self-adhesive glass fibre mesh tape that can be cut with tin snips. It usually comes in rolls of 50 mm × 20 m.
Jointing tape	Used for taping and jointing. A strip of strong, thick, perforated paper with a sticky back that allows jointing compound to enter it. It is pre-creased, to help form internal angles. It comes in rolls of 50 mm × 90 or 150 m.
Rough surface joint tape	Double-sided tape for use on rough or bumpy surfaces.
Scrim tape (Fig 5.46)	This is the most commonly used tape. It is applied before the wall is skimmed with a plaster finishing coat. A flexible strip of mesh or cloth that is sticky on one side. Hessian or jute scrim gives extra strength to wall and ceiling angles. It usually comes in rolls of 50 mm × 90 m.
Wall repair tape	Ultra-thin self-adhesive fabric tape that can be painted over.
Waterproofing tape	To stop water getting into joints in shower and bathroom areas.

Table 5.9 Types of scrim and tape

Jointing compound

Figure 5.44 Corner tape

Figure 5.45 Glass fibre mesh tape

Figure 5.46 Scrim tape

Jointing compound comes in either powdered or pre-mixed form. Many brands are available but generally they are either gypsum-based, which will set fast, or lime-based, which air-dries and need to be applied more thinly and with more coats. You can also buy combined compounds, which are applied in two stages.

As always, follow the manufacturer's instructions when mixing the compounds, and ensure that all your mixing equipment is clean to avoid introducing bumpy particles into the mix.

Sealer

A sealing coat provides a consistent mechanical key over different areas of suction, for example over sanded joints and plain plasterboard. It provides a stable, smooth surface for decorating.

The sealer comes pre-mixed in a can and is applied like paint with a brush or roller.

Finishing dry lining joints

If tapered-edge plasterboard is going to be painted and not skimmed with plaster, you need to seal the joints, which is vital in order to achieve not only a smooth finish but also specified levels of fire resistance and sound insulation. It also reduces movement and cracking around the joints.

To do this, you need to apply scrim or tape and reinforce it with jointing compound. Wait for it to dry and sand it down before you plaster or paint. You should also cover screw holes with jointing compound and sand it down.

You choice of materials will depend on:

* the specification

* the board type

* whether you will use hand tools or a taping gun.

Whatever you use, the **jointing sequence** normally involves several application stages:

1.   bedding the tape

2.   bulk filling the joint

3.   secondary filling to take up shrinkage

4.   finishing by sanding.

After drying, the complete surface is treated with primer or sealer.

See the practical tasks on page 173 for details on how to form and finish joints and angles.

DID YOU KNOW?

Dispensers are cheap to buy and make it much easier to use scrim and tape. Some dispensers are designed to hang from your belt. You can also buy applicators that apply the tape directly to the background but these are much more expensive.

PRACTICAL TIP

First, dust off the surface, especially around the sanded joints, with a brush or broom.

PRACTICAL TIP

Do not overlap tape along the joints or it will be bumpy – apply it side-by-side.

KEY TERMS

Jointing sequence

– taping plasterboard joints and applying plaster to seal them.

DID YOU KNOW?

A taping gun can be used for speed, especially over large areas. You must be trained before attempting to use one.

Maintaining plumb and in-line surfaces

It is important to check plumb as you fix each piece of plasterboard to walls. If you do not maintain plumb (a verticality of less than 5 mm difference from ceiling to floor), you will find that the final board does not fit and you will have to measure and cut it again. This takes time and can result in a poor finish. Boards also need to be kept tight against the ceiling line to prevent gaps.

You just need to place a level on the edge of the board to check for plumb.

Protecting the environment

In Chapter 3, when we looked at sustainability on pages 84–93, we learned that it is important to ensure that all construction work has the least possible negative effect on the surrounding area. One of the ways of reducing its negative impacts is to keep it clean and dispose of waste in an environmentally friendly way. This means recycling as much waste as possible.

It is illegal to send plasterboard and gypsum to landfill, mixed in with other waste. This is because gypsum, when mixed with biodegradable waste, can produce hydrogen sulphide gas in landfill; this is not only toxic but also smells unpleasant. The Environment Agency recommends other steps to take instead:

* Separate gypsum-based material and plasterboard from other wastes on site so it can be either be recycled or reused or otherwise disposed of properly at landfill.

* Separately package or identify plasterboard in a load of mixed waste so that it can easily be identified for separation at a waste transfer station.

* Do not deliberately mix gypsum or plasterboard waste with other waste for landfill.

* Your company's waste management contractor may be able to provide separate skips or sort it for you.

In any case, all waste, no matter what it is, should be sorted and treated before it is sent to landfill.

PRACTICAL TASK

1. FIX PLASTERBOARD USING DOT AND DAB METHOD

OBJECTIVE

To use the dot and dab method to fix 2,400 × 1,200 × 12.5mm plasterboard to a 9m² wall.

PPE

Ensure you select PPE appropriate to the job and site conditions where you are working. Refer to the PPE section of Chapter 1.

TOOLS AND EQUIPMENT

Straight edge	Mixing drill
Spirit level	Buckets
Hop-up	Hammer
Tape measure	Plumb line
Trimming knife	Saw
Foot lift	Pad saw

STEP 1 First, calculate how many plasterboards you will require for this task by dividing the area of the wall by the area of a single sheet of plasterboard.

PRACTICAL TIP

Some plasterers allow for 10% wastage but you should keep waste and offcuts to a minimum as you become more experienced.

STEP 2 Now calculate the amount of bonding compound you require. The manufacturer's instructions on the bag will tell you how many boards or how much area one bag will cover.

STEP 3 Check the wall with a straight edge for any high spots. Remember to allow for the thickness of both the plasterboard and the bonding compound.

PRACTICAL TIP

Remember: always measure twice and cut once when working with plasterboard in order to reduce mistakes and therefore waste.

STEP 4 Measure and mark the finished line of the partition on the floor from the solid background. Allow 10mm for bonding compound and the thickness of plasterboard (in this case, 12.5mm). This gives 23mm.

STEP 5 Transfer the mark by plumbing a line on the wall and then mark the ceiling line from the wall. Check for plumb to the base line mark.

STEP 6 From the internal corner or any opening in the wall, mark the width of the plasterboard (1,200mm) and measure 25mm on each side to allow for the application of the bonding compound. This will stop it from leaking beyond the edge of the boards.

STEP 7 Mark the centres of all the plasterboards. Plumb down from the centres to leave a chalk line that will give the position for the bonding compound.

STEP 8 Cut the plasterboard. Allow 15mm in the height so the foot lift can be used to lift the board in position off the floor.

Figure 5.47 Cutting the plasterboard

STEP 9 Now mix the bonding compound with a heavy duty drill to a creamy consistency.

Figure 5.48 Mixing the adhesive to a creamy consistency

PRACTICAL TIP

Clean the mixing tools straight away. Bonding compound is difficult to remove once it has set.

STEP 10 Now apply dabs of bonding compound to the chalk lines around the perimeter of the first board. Remember to place adhesive along the bottom of the wall and across the middle and top of the wall.

Figure 5.49 Dabbing adhesive on the wall

STEP 11 Gently lift the foot lift until the board in tight against the ceiling. If needed insert packing strips at the base of the plasterboard to wedge it in place, and remove the foot lift.

Figure 5.50 Positioning the boards on wedges

STEP 12 Tap the board firmly using a straight-edge until it lines up with the ceiling and floor lines.

STEP 13 Apply the dabs for the next board and continue the process along the wall. Keep checking the boards horizontally with a straight edge.

Figure 5.51 Checking the boards with a straight edge

Figure 5.52 Boards being pressed in place

PRACTICAL TIP

You may be using tapered edge boards, which means they will not be plastered – so it is important to keep the surface free of adhesive.

PRACTICAL TASK

2. FIX PLASTERBOARD TO A TIMBER JOIST CEILING

OBJECTIVE

To measure, cut and fix plasterboard to ceiling area of 7 m², leaving a background surface ready to receive finishing plaster.

INTRODUCTION

Plasterboards are produced in a number of sizes and types, and can be mechanically fixed to a variety of backgrounds using, nails or screws. For this task you will use basic 1,200 mm × 900 mm plasterboard with a 9.5 mm in thickness.

TOOLS AND EQUIPMENT

Straight edge	Cordless drill
Spirit level	Hammer
Hop-up/scaffold	Saw
Tape measure	Pad saw
Trimming knife	

PPE

Ensure you select PPE appropriate to the job and site conditions where you are working. Refer to the PPE section of Chapter 1.

STEP 1 Erect the scaffold, if you are using it, ensuring there are no traps or gaps and that it is safe and secure.

STEP 2 Check the timber studding ceiling joists are in line, level and at the correct centres, and mark the position of the joists on the wall.

Figure 5.53 Ceiling joists ready to receive plasterboard

STEP 3 Plan the layout of the plasterboards to ensure that the joints are staggered. Measure the ceiling to work out the number of plasterboards required for the task.

STEP 4 Measure for the first plasterboard, working from the left-hand corner of the ceiling, to fix the board horizontally across the joists.

STEP 5 Lay board flat and mark off the measurements required. Then place the straight edge up to these marks and, using the board knife, score the board two or three times along the straight edge. Then turn the board over and tap along the reverse of the cut until the section snaps off. Once the board is broken, cut the paper lining neatly with the board knife.

Figure 5.54 Snapping the plasterboard

STEP 6 Position the plasterboard. Check it fits to the joists then fix it using drywall screws or galvanised clout nails.

PRACTICAL TIP

Consider manual handling guidelines when placing plasterboard above your head. Transport the plasterboard sideways and lift it carefully, preferably with another person.

Figure 5.55 Carrying the plasterboard sideways

Figure 5.56 Lifting the plasterboard with another person

Figure 5.57 Fixing plasterboard with screws

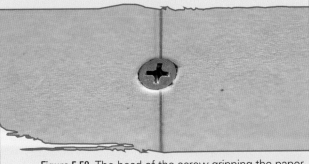

Figure 5.58 The head of the screw gripping the paper without puncturing it

STEP 7 Cut and fix the next plasterboard following the same procedure. Make sure the end joints are staggered to prevent continuous cracking and that a gap of 2–4 mm is left between each board.

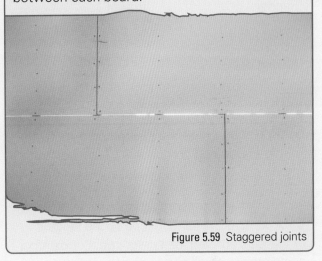

Figure 5.59 Staggered joints

PRACTICAL TASK

3. TAPE AND FILL PLASTERBOARD JOINTS

OBJECTIVE

To tape and fill a plasterboard wall with jointing compound to a smooth finish ready to receive decoration. You will tape and fill an internal angle, external angle and basic joint angle.

INTRODUCTION

When all plasterboards have been fixed to the walls and ceilings, the most common finish is to apply a coat of board finish or Multi-finish ready for decoration.

PPE

Ensure you select PPE appropriate to the job and site conditions where you are working. Refer to the PPE section of Chapter 1.

However, jointing materials also provide a smooth, continuous crack-free surface. You can apply it either manually using hand tools or mechanically using a taping gun.

TOOLS AND EQUIPMENT

Hop-up/scaffold	Jointing tape
Trimming knife	Snips
Sander	Trowel
Broad knife	Sandpaper
Mixing drill	

STEP 1 Make sure the plasterboard joints are free from bonding compound. Clean it off with a sponge or joint rule if necessary.

STEP 2 Cut the jointing tape to the lengths of the plasterboard joints required. For external angle tape you will need to use tin snips.

STEP 3 Check the joint cement manufacturer's specifications to ensure that you only mix the amount required for this task. Mix it to the correct consistency using a mixing drill.

PRACTICAL TIP

Clean all mixing and hand tools before you start to apply the joint cement to the wall.

STEP 4 Apply the jointing tape then cover it with a first coat of joint cement using broad knife or trowel, feathering out each application. Leave it to dry.

PRACTICAL TIP

Some plasterers, like the one in these photos, apply a coat of joint cement before applying the tape. It depends on personal preference, but also the type of tape you use. Self-adhesive tape will not need separate adhesive. Either way, ensure the tape lies flat with no bubbles or folds.

Figure 5.60 Applying the joint cement

Figure 5.61 Applying the tape over the joint cement

Figure 5.62 Flattening the tape

STEP 5 With an angle trowel use the same procedure to cover the internal and external joints. Leave it to dry.

STEP 6 Apply second coat of joint cement. When it is dry, apply the final coat and leave it to dry. The drying time will depend on the environment you are working in – check the manufacturer's specification for guidance. As a rule of thumb, the final coat usually takes 2 hours to dry.

Figure 5.63 Applying a second coat of joint cement

STEP 7 When the jointing material has set you can start to sand down with a basic hand sander or pole sander. You should always wear a dust mask when carrying out this task.

PRACTICAL TIP

When sanding the external angle be careful not to damage the corner.

Figure 5.64 Sanding the filler to a smooth finish

TEST YOURSELF

1. Why is fibre-reinforced gypsum board not classed as plasterboard?

   a. It doesn't have a paper facing

   b. It doesn't contain plaster

   c. It bends too easily

   d. It is not water-resistant

2. What colour is the paper face of fire-resistant plasterboard?

   a. Grey

   b. Pale blue

   c. It does not have a paper face

   d. Pink

3. When setting out for the dot and dab method, what is the measurement you should allow for the thickness of the adhesive?

   a. 10mm

   b. 12.5mm

   c. 22.5mm

   d. 40mm

4. What should you NOT do when stacking plasterboard?

   a. Keep the stack under 900mm high

   b. Stack the boards by type

   c. Lean the boards against a wall

   d. Take it out of its plastic packaging

5. What is the main advantage of direct bond over mechanical fixing?

   a. It doesn't require mixing adhesives

   b. It is better for fixing plasterboard to ceilings

   c. It is suitable for bumpy walls

   d. It is quicker to install and finish

6. Which of these hand tools would you use to cut out holes for sockets?

   a. A pad saw

   b. A rasp

   c. A scoring square

   d. A foot lifter

7. When using the dot and dab method, what percentage of the board area should the dabs cover?

   a. 10%

   b. 20%

   c. 50%

   d. 100%

8. How should joints be arranged across the wall?

   a. In a continuous line

   b. Close together

   c. Staggered

   d. There shouldn't be any joints

9. According to the British Standards for plasterboards, what should be the maximum tolerance for plumb?

   a. ±3mm

   b. ±5mm

   c. ±10mm

   d. None is specified

10. How should you dispose of old plasterboard?

    a. Put it in a skip with other site waste

    b. Bag it up and send it to landfill

    c. Separate it from other waste and recycle it

    d. Burn it

Chapter 6

Unit CSA L2Occ53
APPLY RENDER MATERIALS TO EXTERNAL SURFACES

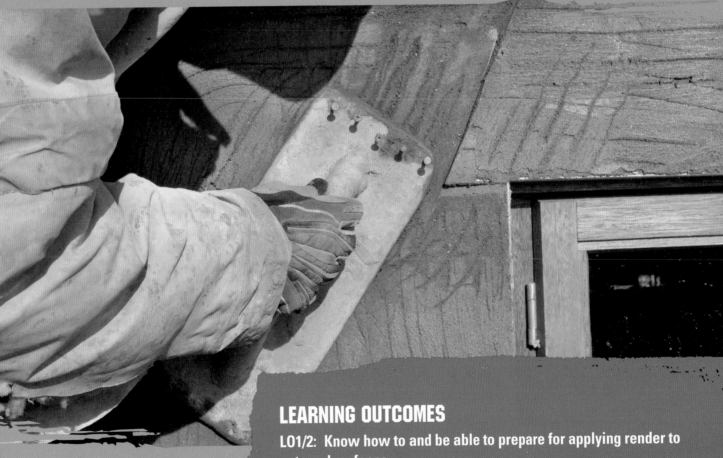

LEARNING OUTCOMES

LO1/2: Know how to and be able to prepare for applying render to external surfaces

LO3/4: Know how to and be able to prepare materials to apply render to external surfaces

LO5/6: Know how to and be able to apply render to external backgrounds

INTRODUCTION

The aims of this chapter are to:

* show you how to prepare background surfaces for rendering

* help you to choose, prepare and mix the appropriate materials for rendering

* show you how to apply trims

* show you how to apply two or more coats of render.

Render is a mortar coating over a **background surface** such as a wall or ceiling. It may be smooth or textured, and it may protect the background surface from damp or extreme temperatures, or just be decorative.

This chapter will show you how to prepare and apply render for external surfaces, such as the outside walls of houses and offices. There are several reasons why a wall might be rendered, such as to:

* strengthen it

* make it weatherproof

* make it look more attractive.

Figure 6.1 Rendering can combine an attractive finish with protective elements

PREPARING TO APPLY RENDER TO EXTERNAL SURFACES

It is important to prepare to render properly in order to get a high quality finish that will not become damp, mouldy or patchy. To get the best result, you will need to consider:

* the suction of the background surface

* the state of the background surface, e.g. whether it is damaged or poorly laid

* the key required for the background surface

* the specified final appearance

* any building regulations or planning requirements

* environmental conditions, such as rain, frost or excessive heat

* access to the wall, e.g. scaffolding

* balancing cost with time and quality.

Hazards and PPE associated with applying render to external surfaces

As with any plastering work, you must always read the risk assessment and follow its recommendations. If you see anything unsafe, report it to your supervisor at once.

Dust and chemicals are hazardous substances, so ensure you wear the correct personal protective equipment. (See the description of different types of PPE in Chapter 1.)

External rendering presents some specific hazards.

Working at height

If you are rendering any wall, you are likely to have to work at height. Look again at Chapter 1 for details of working platforms and access equipment. You should also be aware of the wind picking up while you are working at height – if it is gusty or becoming uncomfortably strong, you should come down and seek advice from your employer or supervisor.

You may also need to wear a harness while you are working at height. Make sure you know how and where to attach this before you start work.

You might also have to transport tools, bags of render or mixed materials to a height. Check if a pulley system is in place, or whether there is another method of bringing up the items you need without any manual handling.

Extreme weather

Rendering also involves working outside, so you will need to protect yourself from the weather. On hot days, you need to cover up your skin when the sun is strong to prevent burning. You might also suffer from heat stress or exhaustion if you spend too long in direct sunshine, which could make you very ill. It is not always possible to get out of the sun at midday, to take more frequent rest breaks or to work in shade, but your employer should have procedures in place on hot days. You should have access to enough drinking water throughout the day.

Freezing temperatures bring their own problems. It is not only difficult and uncomfortable to work outside in the cold, but also bad for your health if you are exposed to freezing conditions for long periods. You should wear several layers of clothing, an insulated jacket, thermal trousers and socks. Insulated gloves are available that still enable you to freely use your hands to work. You should be allowed to take more frequent rest breaks, and to have somewhere you can go to warm up and have a hot drink.

The HSE suggests ways to protect yourself from heat and direct sunshine:

- Keep your top on (ordinary clothing made from close woven fabric, such as a long sleeved workshirt and jeans, stops most UV).

- Wear a hat with a brim or a flap that covers the ears and the back of the neck. Some hard hats have brims and flaps.

- Stay in the shade whenever possible, during your breaks and especially at lunch time.

Figure 6.2 Rendering at height

Figure 6.3 A hard hat with a brim and neck flap for sun protection

Figure 6.4 Dress for the weather

- Use a high factor sunscreen of at least SPF15 on any exposed skin.

- Drink plenty of water to avoid dehydration.

- Check your skin regularly for any unusual moles or spots. See a doctor promptly if you find anything that is changing in shape, size or colour, itching or bleeding.

(*Source* www.hse.gov.uk)

Tools and equipment for applying renders

Many of the tools and equipment you will need are the same as for internal plastering, such as the bucket and brushes in Table 4.1, the hand tools described in Table 4.3, and the bucket, floating and gauging trowels described in Table 4.4. You might also need some of the measuring tools described in Chapter 5, such as levels and measuring tape. Other useful items are:

- an access platform
- a cement mixer
- laser level
- a mortar mill
- a whisk or paddle.

Access platforms may be fixed, such as scaffolding, or mobile, such as a wheeled mini tower or a mobile elevating work platform (MEWP). Never use, move or dismantle them unless you have been trained to do this. While ladders can be used for small jobs taking less than 30 minutes, most rendering jobs take longer than this and a ladder would also not give you adequate and consistent access to the background surface.

The characteristics of rendering materials and components

Render consists of four main elements:

- **sand** (or another **aggregate**), to bulk the mix
- **cement**, to bind the mix
- **plasticiser**, to make it easier to spread
- **water**.

You may need to mix these materials yourself, and include any additives that are needed, or you may use pre-mixed renders, which will be more consistent between batches.

The proportions will depend on what type of finish and material has been specified, as well as the background surface and the coats that are needed.

Sand

The sand you will use for rendering will be quarried, dredged from the sea or rivers, or made artificially by crushing gravel and stones. It is available in a range of colours and the grains should be sharp, well graded, and not bigger than 5 mm. Soft rounded sand, such as the type used in building mortars, is not suitable.

When it arrives on site, the sand may contain silt, which will weaken the mix if it makes up more than 10 per cent of the sand or aggregate's volume and is not removed. You can conduct a simple silt test to find out how much silt is in the sand.

PRACTICAL TASK

1. CARRYING OUT A SILT TEST

OBJECTIVE

To determine the proportion of silt in a quantity of sand.

TOOLS AND EQUIPMENT

Sample of sand or aggregate

Glass container with lid, preferably a graduated cylinder

Water

Salt

Rule or tape measure

STEP 1 Place 50 ml of salt water into the measuring cylinder.

STEP 2 Add the sand sample up to the 100 ml mark.

STEP 3 Pour in more of the salt water solution to bring the liquid level up to the 150 ml mark.

STEP 4 Put the lid on and shake the jar for one minute. Try to level off the aggregate in the last few shakes by using a swirling motion.

STEP 5 Leave the jar to stand for 3 hours, until the aggregate and silt have separated and the water above the aggregate is clear. Then measure the height of the sand and the height of the silt layer.

PRACTICAL TIP

To work out the percentage of silt to sand, you need to use this formula:

$$\frac{\text{Thickness of silt} \times 100}{\text{Total height of aggregate and silt} \times 1}$$

For example, if the silt is 20 mm thick and the total height of the sand and silt is 240 mm, this gives a result of 8.3 per cent. You are looking for a result of 10 per cent or less to be sure your sand is suitable. If it is higher, you will need to wash the sand and test it again before you can use it.

Cement

Cement is a powder that goes through a chemical change or reaction when mixed with water. It becomes like an adhesive paste and then hardens. Its main purpose is to bind bulkier materials like sand and aggregates within the plaster mixture.

In renders, you are most likely to use either ordinary Portland cement or white Portland cement. They both contain 75 per cent limestone and 25 per cent clay, but ordinary Portland cement is grey, and white Portland cement is white.

Ordinary Portland cement is the most common material used when rendering, but many other types are available, such as:

* white Portland cement

* coloured cement

* rapid hardening cement

* masonry cement

* high alumina cement

* sulphate-resisting (anti-salt) varieties.

All cement manufactured in Britain is made to conform to strict British Standards specifications regarding fineness, chemical composition, strength and setting times.

Lime

Lime is crushed and heated limestone, with the chemical name calcium oxide. You will probably only use it on old buildings, as it slows setting time and reduces the strength of the cement. You will come across two types when rendering: hydrated lime and hydraulic lime.

Hydrated lime

This is sometimes called slaked lime, after the process of mixing lime with water to produce calcium hydroxide. It will set, very slowly, in contact with the air. Adding hydrated lime to mortars and renders improves their adhesion and workability and water retention, and it prevents shrinking and cracking. It also helps with suction on the background surface.

If you mix hydrated lime with water and leave it in an airtight container for (ideally) 3 months, it will develop a consistency like cottage cheese, which is known as lime putty. When three parts lime putty is mixed with two parts sand, you can use it as a finishing material. Lime putty is often seen as an environmentally friendly material because it absorbs carbon dioxide as it cures in contact with the air. However, lime putty is slow to set, carbonating at about 1 mm a month.

Hydraulic lime

Although it is often confused with hydrated lime, hydraulic lime is quite different because it will set when slaked (mixed with water). This is because it contains impurities that react to harden the render. Some suppliers make a distinction between natural hydraulic lime, which contains natural impurities like clay or silicates, and standard hydraulic lime, which contains materials like cement, blast furnace slag and limestone filler. Because it is slow to set compared with cement-based materials, and remains soft underneath, it should only be used where breathability is required and strength of the render is not important.

Additives

Without additives, cement can make plaster and render shrink and crack as it dries. Too much water in the mix will make the render weak. In hot weather the mix can dry too quickly and in cold weather it can dry too slowly or be damaged by frost. Standard mixes are not waterproof and most mixes will develop surface pores as air bubbles escape while the mixture dries, causing weakness and allowing water penetration.

Additives are designed to make working with plaster and render easier and produce better results. Additives may have been added to pre-mixed render when it was manufactured or you may need to add them yourself.

Look back at Table 4.7 on page 116 which describes the main additives that you might use in both external render and internal plaster. They may be in the form of liquids or powders, and some products combine more than one property, for example a waterproofer and retarder.

Figure 6.5 The Great Hall at Stirling Castle in Stirling, Scotland

PREPARING MATERIALS TO APPLY RENDER TO EXTERNAL SURFACES

Using well-graded sand

As we saw on page 181, it is important to use the right sort of sand in your render. Whether you are using a cement or lime-based render, the sand must be clean, sharp and well-graded.

Sharp means that the grains are rough and irregular in shape, not smooth, slippery and round like builders' sand. Confusingly, the finer gradings of sharp sand, like those used in plastering, are technically still sharp but are often called soft sand, because they do not contain visibly separate pieces of grit. They may also be called plastering sands.

Well-graded means that there is a mixture of small, medium and large grains in the sand. If you put a sample through a set of grading sieves, some will be left in each size of sieve mesh. Grains of different sizes mean that there is less space between the grains, making the sand stronger and denser. Without smaller grains to fill the spaces or voids, more cement or lime would be needed to strengthen the mix.

Sand and aggregates are dug from a pit or river bed, or are quarried, so are natural materials. Although sand from different sources may look the same, the quality of different batches may vary.

PRACTICAL TIP

Although sharp sand is generally best for the scratch coat, a 'rendering sand' based on a mix of building sand and sharp sand is often used for the topcoat. Check the specification to find out what is required.

Badly graded sand: voids or spaces between the grains

Well-graded sand: voids filled with medium and small grains

Figure 6.6 Badly and well-graded sand

Sea sand is often thought to be unsuitable because of the risk of **efflorescence** due to the presence of salt. Some plasterers, however, believe that salt in the air or alkali cement is more of a problem, especially if the sand is washed before use.

Advantages and disadvantages of using lime

As we have seen earlier in this chapter, lime is preferred for use in older buildings that need to 'breathe'. It also improves the properties of the render, such as its:

* flexibility (by offsetting movement caused by traffic vibration or the foundations settling)
* workability (as it retains water long enough for the mix to be worked)
* plasticity (by providing good adhesion to the background surface)
* anti-fungal properties (as the alkali in lime is naturally fungi-repellent).

Cement renders are inflexible and not porous, as lime is, so may crack over time. If water gets trapped in the crack, the wall will become damp and damaged.

Other advantages of lime are related to the environment:

* Lime renders are easy to remove from walls, so the bricks can be reused.
* The production of lime produces 20 per cent less carbon dioxide than cement production.
* Lime is biodegradable.
* Lime is recyclable.

Figure 6.7 Sharp sand ready for use

Figure 6.8 Grading sieves

However, the setting process of lime plasters and renders is not as standard and predictable as it is for cement- and gypsum-based plasters and renders. Lime materials take a long time to dry, which is often impractical when a site has a schedule to stick to, and trades are booked in for particular times.

Lime is also significantly weaker than cement, so should not be used when strength is an important specification – but it is preferred for older and heritage buildings so make sure you check the specification carefully.

Using good quality water

Always use cold, clean water in any render mix. Just as we have seen how impurities in the dry materials can affect the render, dirty water is likely to weaken the mix or cause staining on the finish.

Use the minimum quantity of water to ensure workability – it is easier to add more water than to try to remove it or risking spoiling your mix proportions by adding more sand or cement.

Figure 6.9 Efflorescence on a rendered wall

Trims and beads

As we saw in Chapter 4 (pages 109–111), beading or trims are used internally and externally as edgings to get a sharp corner, for example where the wall meets the ceiling. Externally, they are also used for stops and **bellcasts**. Stainless steel beading is used for external work, or where there is a high moisture content, and galvanised steel is used for internal work, although uPVC beading can be used both internally and externally, and is also available in different colours.

Beading strips come in different sizes, and are cut to size with tin snips or a hacksaw or nailed together. They can then be bedded into the scratch coat.

Beads have a number of benefits:

* They strengthen corners and edges, so that the plaster is less likely to be chipped.

* Ready-made edges mean that features like arrises, stops and movement joints don't need to be made by hand (see page 113).

Table 6.1 describes different types of beads that you may use when preparing to render.

Type of bead	Use
Angle bead (Fig 6.10)	To form external angles and prevent chipping or cracking. These come in different sizes and angles.
Bellcast bead (Fig 6.11)	Also called a render stop bead. This forms and protects the lower edge of external render. Designed to deliver a gentle gradient at the base of the render, it is used above doors, windows and at DPC level to allow rainwater to drain clear of the background surface.
Drip bead	To be used flush above doors, windows and at damp course level. It prevents the retention of water that often stains the bottom of the render.
Movement bead (Fig 6.12)	Also called an expansion bead. This allows + or – 3 mm of movement between adjoining surface finishes or sections that might move. They can also be used where changes in the render colour are specified.
Stop bead (Fig 6.13)	To provide neat edges to two coat plaster or render work at openings or abutments onto other wall or ceiling surfaces. Render (external) versions help with water run-off.

Table 6.1 Types of beads and their uses

KEY TERMS

Bellcast

– a curve at the bottom edge of a roof or external wall, formed with or without beads, to divert rain away from openings or bases of walls.

DID YOU KNOW?

BS EN 13139 is the specification for aggregates for mortar. This specification covers sands for masonry mortar, plastering mortar, rendering mortar, floor screed, special bedding materials, repair mortars and grouts. The sand you use should state on the packaging that it complies with this specification.

PRACTICAL TIP

Tap water is fine to use – just ensure that you use a clean bucket to hold it.

PRACTICAL TIP

Don't use galvanised steel trims for external work as they will go rusty.

Figure 6.10 Figure 6.11  Figure 6.12  Figure 6.13

Establishing the correct mix proportions

It is important to ensure you mix the correct proportions of each material – otherwise your render will be too wet, too dry, will crack or be the wrong strength.

Render materials are traditionally measured by volume. You can use any suitable container to gauge the amount you need, such as a bucket, bag or wheelbarrow. Do not use a shovel as it is hard to ensure accurate and consistent batching.

You can also weigh the materials, but batching them in buckets by volume is much easier.

Use the same method for different materials and batches. As you will normally need to mix a fairly large quantity to cover the wall, a cement mixer is more efficient than attempting to mix it by hand or with a whisk, and will give a more uniform mix.

Order of the mixing process
1. Pour water in first.

2. Add some sand.

3. Add the cement.

4. Add the remaining sand.

5. Pour in more water.

6. Add an additive if required.

7. Allow the materials to mix for 3 minutes.

Proportions
The proportions depend on the materials you are using and the suction of the background. A starting point is four parts sand to one part cement for the scratch coat, with more sand in later layers. Include additives at the ratio specified on the packaging, ensuring that they can be used with the type of cement you are using.

Pre-mixed renders have already been gauged to the correct strength – you just need to add the amount of water specified on the packaging or manufacturer's data sheet.

APPLYING RENDER TO EXTERNAL BACKGROUNDS

There are several steps to ensuring a successful rendering job. You need to:

* understand the nature of the background surface

* prepare the background

* select and mix the most appropriate render

* apply a suitable scratch coat

* cure the scratch coat

* apply a suitable topcoat or finishing coat

* cure the topcoat.

The importance of compatibility between background surfaces and render to be applied

Background surface suction

As with internal plastering, different types of background surfaces have low, medium or high suction.

Low suction

Backgrounds like painted surfaces and pre-cast concrete are not very porous. The fresh render is likely to sounds hollow when you tap it or even fall off the wall because it will set and cure on top of the surface as there is not enough suction for it to stick.

Figure 6.14 A low suction external background

Medium suction

Backgrounds like engineering bricks, common bricks and medium-density blocks are slightly porous and need little preparation. The render will firm up by setting naturally or by evaporation in warmer areas so the background may still need a bonding coat.

High suction

Backgrounds like some types of brick, aircrete blocks and old render are porous, so they will suck moisture from the fresh render and dry it out too fast. This makes the render too difficult to work with as it may be dry before you can make it smooth, or it will crack after you have finished rendering the wall.

Figure 6.15 A medium suction external background

Preparing the background

A newly constructed medium suction background (such as the wall of a new house) will just need damping down with water before rendering, while low and high suction backgrounds, and those that have previously been rendered, or are damaged, will need additional preparation.

All surfaces should be brushed down to remove dust, dirt and lichen, and a fungicidal coating should be applied if the bonding adhesive or render do not contain one.

* **New blockwork**: brush and damp down

* **New brickwork and concrete**: brush and damp down then apply a slurry bonding adhesive (spatterdash coat, see Chapter 4) to form a key.

Figure 6.16 A high suction external background

* **Old brickwork and stonework**: hack off the old render and rake out the joints before damping down and applying the spatterdash coat. You may also have to remove bump and ridges, and fill in holes and other damage with mortar.

Bonding adhesives

A bonding adhesive improves the adhesion of the scratch coat. In addition to the traditional spatterdash coat, which you can make yourself, other pre-mixed slurries and adhesives are available:

* **Pre-mixed bonding adhesives**: usually a combination of sand, other lightweight aggregates, hydrated lime, white cement and additives to improve workability, adhesion and reinforcement.

* **Polymer-modified slurries:** these are pre-mixed with cement and other materials. You just need to add water.

* **SBR** (styrene butadiene rubber) is a latex-based liquid that is particularly effective for external work as it is water-resistant. It may be used on its own or mixed with cement or water.

To apply the bonding agent, follow the instructions on the packaging – you will need to dilute it with water before applying it to the wall and it may be necessary to add sand to provide a rougher texture. Several coats may be needed before the surface is sealed and you will have to wait at least 12 hours before the wall is ready to render but you should also ensure it is still tacky.

Remember that the bonding adhesive must be compatible with the render you are going to use, and that the render itself has to be appropriate to the background surface – for example, it may need to be porous so should contain lime. Make sure you read the manufacturer's data sheet before starting work.

Selecting the right render

The type of render you use should be appropriate to the background surface. As a general rule, it should not be stronger than the surface to which it has been applied, otherwise it may pull the wall down. Table 6.2 describes which renders would suit which backgrounds. Note that this refers to a particular manufacturer's rendering products, so may not apply to all brands of pre-mixed render or types of cement.

> **DID YOU KNOW?**
>
> PVA is generally unsuitable for external work because it is water-soluble. However, some brands produce external PVAs – check you have the right type.

> **PRACTICAL TIP**
>
> Generally, you'll need a stiff mix for blockwork and a slightly wetter mix for brickwork.

Render designation	Render characteristics	Typical substrates
i	Strong, relatively impermeable with high drying shrinkage	Engineering bricks, in situ concrete, dense concrete blocks
iii	Moderately strong	Calcium silicate bricks, some facing bricks
iii	Medium strength with greater permeability than Designation i, but less likely to crack	Lightweight aggregate blocks, some common bricks, aerated concrete bricks
iv	Moderately low strength	Aerated concrete blocks, some softer bricks
v	Low strength	Weak materials in sheltered locations

Table 6.2 Typical suitability of different render designations (*Source* Lafarge Cement)

Applying two or more coats of render

When you have prepared the background, you are ready to apply the render, normally as two or three coats.

* **Two coat work** consists of a scratch coat and a top coat and is used if a flat (plain faced) render is specified.

* **Three coat work** has an additional coat that forms a decorative finish, like a pebble-dash. A third coat might also be a dubbing out coat to fill holes and flatten the surface before the scratch coat is applied.

The same type of cement should be used in both the scratch coat and the top coat.

The scratch coat

As with internal plastering, this is the first coat that is applied, to control suction, straighten and even out walls and provide a mechanical key for next coat. It is thicker than the top coat – 9 to 12 mm – to ensure that the surface is as smooth as possible.

The mix should be about four parts sand to one part cement and one part water, and normally includes a waterproofer.

After it has dried for 15 to 20 minutes, scratch the surface in shallow horizontal waves to provide a key for the top coat.

The practical task on page 193 describes how to apply a scratch coat.

The top coat

The top coat provides a smooth finish. It is applied about 10 mm thick and smoothed with a straight edge or darby rule followed by a float. Pre-mixed top coats are available, which give a more attractive final texture, or contain coloured pigments. When pebble-dashing, the top coat is known as a **butter coat**, and the aggregates are applied at the same time.

If you are using cement-based render, you should wait at least 24 hours before applying the top coat to allow the scratch coat to set and cure. Setting time for lime-based render is much longer, and you are likely to have to wait several weeks. However, you should not leave the scratch coat exposed to the air too long before applying the top coat, or it will shrink and crack.

The decorative coat

If there is a decorative coat it is also usually pre-mixed to provide texture or colour. This coat might consist of aggregates being embedded in the soft top coat, as in the case of pebble-dashing. It is important to ensure that the same colour or texture is applied evenly across the whole wall.

After the final coat has been applied, you can cover the render with a damp canvas or tarpaulin, to help it to cure and also to protect it from the weather and items in the air becoming stuck to it.

PRACTICAL TIP

These days, machines may apply the bulk of the render to the wall but you still need to finish each coat yourself.

PRACTICAL TIP

Remember to provide a key for the next coat of render. This may be mechanical (by scratching into the surface) or by applying a spatterdash coat or bonding adhesive.

KEY TERMS

Butter coat

– the soft final coat to which the aggregate is applied in dry dashing.

PRACTICAL TIP

Measure the amount of materials correctly. Where the top coat has begun to dry quickly, you can apply water before the rubbing up process begins. Be careful not to apply too much water directly onto the finished surface as this may remove some of the cement in the mix, creating a sandy, weak and possibly blotchy finish on the surface. The timing of the rubbing up process is very important, as are the weather conditions at the time of applying the materials.

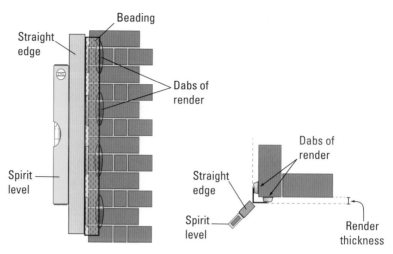

Figure 6.18 Applying beading to an external wall using render

Figure 6.17 A decorative finish

Fixing beads and trims

Before you start your scratch coat, you need to fix beads and trims to external corners and door and window reveals. These provide a sharp edge and form stops and bellcasts.

Beading and trims usually come in 2,400 mm and 3,000 mm lengths. Measure the length required and cut it to size with a small hacksaw or tin snips.

You can fix beading or mesh in place with dabs of render and/or secure them with mechanical fixings like steel clout nails, rustproof screws or staples.

Dab small amounts of render at regular intervals onto the wall along the length you need. Then gently apply the beading and tap it into place. Once you have checked for plumb and level, cover the wings in render, wiping off any surplus render before it dries. If you do not bed it in at this stage, the adhesion may be reduced because it is difficult to squeeze plaster between the bead or mesh and the background surface.

Fixing bellcast beads

Set bellcast beads in a continuous line across the wall above windows and doors to prevent rainwater reaching the damp-proof course (level). Their job is to encourage rainwater to run off so any gaps will make it run down the wall instead. You can also fix the bellcast beads by hammering in plug fixings every 300 to 700 mm.

See page 192 for a practical task on how to fix a bellcast.

Fixing stop beads

Stop beads are placed around door and window reveals, or at a change of finish, to prevent cracking. Like bellcast beads, they can be fixed either mechanically with nails at intervals of 150 mm or onto dabs of render or mortar.

Using timber battens

If you are restoring a historical building, you may find that timber battens (sometimes called arris sticks) are more appropriate than stainless steel or uPVC beading. These are temporarily fixed around corners and the render applied inside them. They are then slid away when the render is nearly dry. Some experienced plasterers prefer this method because:

* it looks better on old buildings than modern trims, which may appear to provide an unnatural angle, or show through the render

* if the corner is damaged, the whole length of a buckled trim would need to be replaced, whereas a small chip on an un-beaded corner is easier to repair

* a bead may present a weak area where damp can enter the building.

See page 195 for a practical task on forming angles without using beading.

CASE STUDY

South Tyneside Homes

South Tyneside Council's Housing Company

Plastering for a good cause

Dave is the plastering tutor at Vision West Nottinghamshire College.

'While I was teaching at my old college, I found out that my wife's best friend had got breast cancer. The learners asked what they could do to help and we came up with the idea of a sponsored 24-hour "plasterthon". Along with my colleagues, Sam and Chris, I went round skimming all the bays in the college workshop and 10 apprentices followed us, taking it off again. I plastered each bay four times in 24 hours – it was about the area of a football pitch.

It was a great atmosphere and everyone got involved – we were in the local press and the local burger van brought us breakfast cobs in the morning. Best of all, we raised £1,450. But my arm ached for days afterwards!'

1. FIX EXTERNAL BELLCAST AND EXPANSION BEAD

OBJECTIVE

To fix an external bellcast and expansion bead (2,400 mm in length) plumb and ready to receive a scratch coat of sand/cement render.

INTRODUCTION

Bellcast beads or render stop beads are used over window and door openings and above damp-proof courses for external renders. Expansion beads are used to bridge two surfaces at a gap or joint that is likely to expand and prevent the rendering from cracking.

TOOLS AND EQUIPMENT

Plastering trowel	Straight edge	Spot board
	Spirit level	Hop-up
Handboard/ hawk	Gauger	Buckets
Beads	Flat brush	Drill

PPE

Ensure you select PPE appropriate to the job and site conditions where you are working. Refer to the PPE section of Chapter 1.

STEP 1 Measure the length required and cut the beads carefully with a pair of tin snips. Always cut the mesh side first and wear protective gloves.

STEP 2 Place the bellcast bead horizontally two courses from the floor or above the damp-proof course. Mark its position and then put it aside.

STEP 3 Apply sand/cement mortar along the area in continuous dabs then position the bellcast for secondary fixing. Either hammer steel clout nails into the mortar joints or fix plugs and steel screws to support the bead.

STEP 4 Check the bead for level and tighten in the mesh with a coat of sand and cement to strengthen the bead.

PRACTICAL TIP

Always apply sand and cement to the full length of the bead leaving no mesh exposed.

STEP 5 Brush the bead with a flat brush to clean off any excess mortar.

Figure 6.19 The bellcast bead in position

PRACTICAL TIP

Follow the same procedure to fix the expansion bead but make sure that the bead overlaps the joint or change in backgrounds and it is vertically plumb. When it is in place, clean out the middle of the expansion bead with a flat brush and make sure that the thickness of render to be applied over it will be no more than 12 mm.

Figure 6.20 The expansion bead in position

PRACTICAL TASK

2. APPLY A PLAIN-FACED RENDER COAT

OBJECTIVE

To produce a plain-faced rendered wall that would be level, flat and accurate to within ±3 mm in a 1.8 metre length to an area of 7 m².

INTRODUCTION

Plain-faced render is the most common external render finish that you will come across. The golden rule is to make sure that the first coat (scratch coat) is a stronger mix than the second coat. This is because strong mixes tend to shrink, and this imposes stress on the scratch coat that may result in cracking and bond failure. The second coat should be applied to a thickness of 6–8 mm, depending on the background.

PPE

Ensure you select PPE appropriate to the job and site conditions where you are working. Refer to the PPE section of Chapter 1.

PRACTICAL TIP

The render should be finished using a plastic or wooden float. It is poor practice to finish the wall with a steel trowel.

TOOLS AND EQUIPMENT

Plastering trowel	Flat brush
Handboard/hawk	Spot board
Straight edge	Hop-up
Darby	Buckets
Spirit level	Scratcher
Gauger	Float

Figure 6.21 Trowel, mortar, float and scratcher, ready for use

STEP 1 Prepare the background before you start. Clean off any mortar and, if required, apply a coat of PVA to control the suction.

PRACTICAL TIP

If you are practising applying render at college, you will probably use a bay inside the building. However, if you are applying the sand and cement to an external wall you need to apply SBR to control the suction. PVA is water-soluble so it is not suitable for using outside.

STEP 2 Mix the sand and cement to a ratio of four parts sand to one part cement. For example, a four to one mix would require four level buckets of sand to one level bucket of cement. Add water proofer and clean water.

STEP 3 Apply the scratch coat, working from the top right-hand side of the wall to a thickness of approximately 8–10 mm.

STEP 4 Allow the first coat to set for about 20 minutes then form a key by making wavy horizontal lines with a scratcher.

Figure 6.22 Keying the scratch coat

STEP 5 After the scratch coat has been left to dry for 24 hours, mix the material for the second coat at a ratio of five parts sand to one part cement. Do not add any waterproofer to the second coat mix.

STEP 6 Apply the second coat of sand and cement following the same process as you did for the scratch coat (Step 3) but also rule it off with a straight edge and fill in any hollows.

Figure 6.23 Applying the second coat

STEP 7 Pass a float over the surface using small circular movements, with minimal pressure, until you achieve a blemish-free flat surface.

Figure 6.24 Rubbing up the surface

Figure 6.25 The rubbed up surface, compared with the ruled off second coat

PRACTICAL TASK

3. FORM AN EXTERNAL ANGLE WITHOUT THE USE OF BEADS

OBJECTIVE

To form an external angle using sand and cement, achieving a smooth finish that is level and straight to ±3mm in a 1.8m length.

INTRODUCTION

On some traditional work, or where matching existing work is necessary, beads are not suitable for external rendering and the plasterer has to form the angle using sand and cement and carrying out the reverse rule method.

The process is the same as that for forming external angles with plaster – see Practical Task 7 in Chapter 4.

PPE

Ensure you select PPE appropriate to the job and site conditions where you are working. Refer to the PPE section of Chapter 1.

TOOLS AND EQUIPMENT

Plastering trowel	Flat brush
Handboard/hawk	Spot board
Straight edge	Hop-up
Darby	Buckets
Spirit level	Scratcher
Gauger	Float

PRACTICAL TIP

This task is best completed by two plasterers.

STEP 1 Follow steps 1 to 5 of Practical Task 2 above.

STEP 2 Place the rule on one side of the angle and apply the render floating up to the rule.

STEP 3 Then carefully remove the rule and place it on the previously floated wall. Float up to the rule and carefully slide it away, leaving a square corner ready for rubbing up.

STEP 4 Carefully rub up the external angle, working away from the corner and fill in any hollows as you go.

TEST YOURSELF

1. What is a disadvantage of using lime?

   a. It sets so quickly that it is difficult to work

   b. It makes the wall less porous

   c. It is difficult to achieve a smooth finish

   d. It takes much longer to set than cement

2. Which of these is an example of a high suction background?

   a. Pre-cast concrete

   b. Aircrete blocks

   c. Paint

   d. Engineering bricks

3. What is another term for frostproofer?

   a. Accelerator

   b. Plasticiser

   c. Retarder

   d. Expanded perlite

4. Why would you use a bellcast bead?

   a. To form an external angle

   b. To direct rainwater away from the wall

   c. To allow for movement between sections of the wall

   d. To provide a mechanical key for the top coat

5. What effect will adding more sand have on the render?

   a. It will make it waterproof

   b. It will strengthen it

   c. It will weaken it

   d. It will overplasticise the mix

6. Which of these is often described as a slurry?

   a. SBR

   b. Lime putty

   c. Sand mixed with water

   d. The spatterdash coat

7. What is efflorescence?

   a. A white coating of salt crystals

   b. The ability of the render to reflect light

   c. A render that dries very quicky

   d. A mineral added to the render mix to improve its fire-proofing qualities

8. What is the best water to use in a render?

   a. Lukewarm and clean

   b. Cold and clean

   c. Warm mineral water

   d. Water reused from a previous mix

9. What is the best way to cure a render?

   a. Leave it to dry naturally in the air

   b. Cover it with damp canvas or tarpaulin

   c. Blow warm air at it

   d. Spray it with a finisher

10. What is a dubbing out coat?

    a. A decorative coat applied last

    b. Another name for a scratch coat

    c. A coat applied to fill holes in the background surface

    d. A coat of SBR to act as a bonding agent

Chapter 7

Unit CSA–L2Occ55
FORM SAND AND CEMENT SCREEDS

LEARNING OUTCOMES

LO1/2: Know how to and be able to prepare for forming sand and cement screeds

LO3/4: Know how to and be able to prepare materials for forming sand and cement screeds

LO5/6: Know how to lay sand and cement screeds

INTRODUCTION

The aims of this chapter are to:

* describe the tools, equipment and materials you need to lay floor screeds

* explain the different types of floor screeds and when you would use them

* show you how to set out datum marks for levels and falls

* help you to mix screed material correctly

* show you ways of laying floor screeds to levels and falls

* explain how to compact, finish, cure and dry new floor screeds.

LAYING A FLOOR SCREED

KEY TERMS

Floor screed

– a layer or layers of screed material laid in situ, usually over a concrete sub-base to obtain a defined level, to carry the final flooring or to provide a wearing surface.

Although there are specialist floor screeders, it is often the plasterer's job to lay a semi-dry **floor screed** after the walls and ceiling have been plastered. The purpose is to create a floor surface that is flat and smooth enough to be finished with tiles, carpet, laminate, floorboards or another floor covering. If the floor is not flat enough, the floor covering will not lie flat, which not only looks unattractive but could also cause people to trip or furniture to tip. It's therefore important to take time and care to get screeding right.

KNOW HOW TO PREPARE FOR FORMING SAND AND CEMENT SCREEDS

The hazards of forming sand and cement screeds

In Chapter 1 we examined the importance of identifying potential hazards and the ways in which risk assessments and method statements can help to avoid possible health and safety hazards.

The most common hazards you are likely to encounter when laying floor screeds are:

* screeding materials heating up when they are mixed, which could burn bare skin

* kneeling, crouching or bending for too long as you lay the floor, which could lead to back and knee problems

* poorly maintained and/or wrongly used power tools, which can result in injury

* on large sites, being hit by travelling plant or by something falling on your head

* poor housekeeping, for example tools being left around or cables crossing a room, resulting in slips, trips or falls

* injuries associated with manual handling.

It is also important to make sure that manufacturers' instructions are followed. These precautions ensure that you follow health and safety legislation and keep yourself safe.

PPE and protecting yourself from injury

Cement can burn your skin so always make sure that you cover up when working with it. Wear waterproof trousers instead of shorts, and long sleeves instead of T-shirts. Always wear gloves, goggles and overalls, and a face mask if you are dealing with fine, dusty particles. Wear a hard hat if necessary.

Preparing a floor involves a lot of bending and kneeling, which can put stress on your joints and result in long-term injuries to your knees and back. Kneepads or a knee mat distribute weight and pressure away from the knees, making it more comfortable to maintain a kneeling position. However, you still should get up regularly to walk or stretch.

Useful information sources about sand and cement screeds

As with any type of plastering, you must follow the specification provided for the job, along with any other information provided to you. Pay particular attention to drawings that show, for example, the required height of the floor. It is also important to note the type of floor covering (carpet, laminate, tiles etc.) that will be used on the screed, as this will affect your finish.

Several **British Standards** and European standards apply to the laying of floor screeds and the materials used.

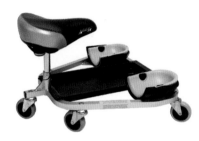

Figure 7.1 A kneeling creeper

KEY TERMS

British Standards

– the British Standards Institute (BSI) develops and publishes standards in the UK. These standards are not usually legally binding but they represent best practice when carrying out different activities. British Standards are usually a number starting with BS, or BS EN if they are also a European standard.

PRACTICAL TIP

Always check the manufacturer's data sheet and the guidance printed on the bag of screed before you mix it. Floor screeds require a lot of material so, if you get it wrong, that's a lot of money wasted.

These include:

Code of standard	Name of standard	Contents of standard
BS 8000-9:2003	Workmanship on building sites. Cementitious levelling and wearing screeds. Code of practice.	This standard provides recommendations and gives guidance on laying cementitious levelling screeds and wearing screeds, including on the quality of materials to be used.
BS 8204-1:2003+Amendment 1:2009	Screeds, bases and in situ floorings. Concrete bases and cementitious levelling screeds to receive floorings. Code of practice	This standard gives recommendations for the design and laying of concrete bases and cement levelling screeds to receive flooring.
BS EN 13318	Screed materials and floor screeds – Definitions	This standard provides agreed definitions for the technical terms being used in BS EN 13813, in English, French and German, so that they are clearly understood across the whole of Europe.
BS EN 13813	Screed materials and floor screeds – Screed material, Properties and requirements	This is the product standard that describes the essential characteristics of flooring products, specifies the methods by which these characteristics are to be determined and defines minimum levels of acceptable performance.
BS EN 13892	Methods of test for screed materials	A suite of eight test method standards for things like wear resistance and surface hardness.

Table 7.1 British and European standards relating to floor screeding

Tools and equipment for floor screeding

Some of the tools and equipment you will use will already be familiar from other plastering tasks. Others are used especially for floor screeding. Table 7.2 describes the main ones.

Tool/equipment	Description
Bucket	Buckets need to be big enough to contain the volume of your mix, and completely clean before you use them.
Builder's square	This is used to set out right angles – in floor screeding, this is when you are positioning the screed from the data point. It is wooden and braced with one side longer than the other. Squares are made from 75 mm × 30 mm timber and are half jointed at the 90° angle with a diagonal brace. Modern builder's squares are also often made from aluminium and can fold up.
Cement pump	This is used to transport large quantities of screed to where it is required. For example, if your screed is mixed away from where it is going to be laid, such as outside the building, a cement pump makes it easier to move it to where it is needed. The screed needs to be wet enough to flow easily through the hose. As with any electric equipment, you should not use a cement pump unless you have been trained to do so.
Chalk line	This is used to transfer floor thickness datum points around the area to be screeded. It is a reel of string into which you pour coloured powdered chalk, so that it transfers a straight line onto the surface as you reel it out. Some models can also be used as a plumb bob – the heavy end is dangled from the string to identify a vertical line.

Tool/equipment	Description
Float	This can be made out of wood or plastic and is used for smoothing and filling in hollows after you have ruled in the screed.
Floor laying trowel	This is used for laying, mixing and spreading flooring materials like sand and cement, and for giving the screed a smooth finish. It is made from steel and is either 400 mm or 455 mm long, has a pointed front and is normally much larger than the plastering trowel.
Gauging trowel	This is used to mix small amounts of material and to get plaster into difficult places, like corners of floor screeds. It can also be used to clean down other tools after screeding. It is made of steel and usually has a wooden or plastic handle.
Measuring tools	These include a long tape measure and a steel rule for measuring the thickness of the screed.
Mixer	A self-contained mixer is useful for mixing screed on larger jobs. Portable cement mixers have a capacity of 60 to 150 litres. They are powered by electricity (110V or 230V) or petrol.
Shovel	Shovels must be the right size for your height and strength. This is because you can easily strain your back when working with a loaded shovel.
Straight edge	This is a long, straight piece of wood or aluminium for levelling in the screeds and for ruling them off. They are usually about 3 m long, although they are available in shorter lengths, and about 100 mm wide and 35 mm deep. It's best to buy one rather than try to make one yourself, as it's important that the edges are completely parallel and the faces of the depth are flat.
Wheelbarrow	Wheelbarrows reduce the time and strain involved in transporting many heavy bags or loose sand or cement. This is essential when screeding which uses up a lot of material. Choose a wheelbarrow that is the right size for the job you are doing, e.g. ensure it will fit through doors and go around corners. Try to keep it clean so that the materials don't get contaminated.

Table 7.2 Tools and equipment used for floor screeding

You will also need to use levels to ensure that the floor is flat and to mark datum points. Table 7.3 describes the main types you will come across on site.

Type of level	Description
Laser level	This level sends out a laser beam to a target receiver. The level at that point can then be recorded for setting out. Models range from large industrial to small auto-setting versions. (See Fig 7.11.)
Automatic liquid level	This consists of a container of coloured liquid which is placed at a high point and attached with a hose to a metal gauge with an air valve. The datum level is set on the scale with a locking screw and water is drawn up to that level. The datum level can then be marked on the wall. (See Fig 7.12.)
Spirit level	This is a simple straight edge that has a glass tube containing a liquid and a bubble of air. When the air bubble is in the centre of the tube, the straight edge is exactly level. The tubes are marked with lines to confirm that the edge is level.
Water level	This is a length of hose that has a transparent tube in each end. The hose is filled with water. It is an ideal resource for checking levels in distances of over 30 m. The water level is a simple device, as water will always find its own level. One of its advantages is that you can take it around corners or obstructions. Note that these days it is not often used for levelling in new buildings.

Table 7.3 Types of level

Figure 7.2 Ordinary Portland cement

Figure 7.3 Sand suitable for floor screed

PRACTICAL TIP

Always clean your tools and equipment after you have used them – it will save you time in the long run as dried screed is very difficult to remove. You should also regularly check that your squares and levels are still accurate. Replace any tools that are damaged.

Materials and resources used to form sand and cement screeds

Materials for floor screeding usually contain 1 part ordinary Portland cement to 3 or 4 parts sand/aggregates. You can also get ready-mixed screed materials, often containing retarders to slow the set. This means more time can be spent laying the screed and less time spent shovelling sand and cement into the mixer.

Cement

Cement is added to bind the mix – that is, to stop it from falling apart. It is important to choose the specified cement type, because variations in cement quality can affect the strength and curing of the screed. The type of cement used for floor screeds is called **ordinary Portland cement**, which comes in different types according to the type of screed you need. Normally, you would use Portland Class 42.5N, to produce a fine concrete screed.

The initial set of a mix using ordinary Portland cement should not be less than 45 minutes and the final set not more than 10 hours.

Sand

The size and shape of aggregates affects how well the screed performs. The aggregate type specified for ordinary cement-sand levelling screed is washed sharp sand with particles of up to 4mm, as this makes it more hard wearing. Sand used for screeds should always be clean and well graded as it makes the mix more workable.

It is important to get the amount of sand in the mix right. Sand is cheaper than cement so it may be tempting to put in as much as possible – but the more sand in the mix, the weaker its strength, making it difficult to work. However, not enough sand (an under-sanded mix) will dry too quickly and the screed is likely to crack.

For general fine concrete screeds, you can replace 25 per cent of the sand with a single-sized aggregate, like granite chippings, using a mix of 1 part cement, 1 part aggregate and 1 part sand.

Do not use single-sized finely graded sand, fine brick-laying sands, crushed rock and sea sand containing shell.

Water

As when mixing any plaster, use enough clean, running water to work easily with your mix.

Polypropylene fibres

You can reinforce the screed and prevent it from cracking by adding polypropylene (PP) fibres. Beware that PVA (polyvinyl acetate) fibres are sensitive to water so should not be used in damp areas.

Admixtures

Admixtures can be added to ordinary Portland cement for different purposes, including:

* rapid-hardening
* retarding
* sulphate-resisting

* heat-resisting
* super-plasticising
* waterproofing/water repelling.

You may also add pigments to give it a particular colour.

Damp-proof membranes

To comply with Building Regulations, all new buildings must have a damp-proof membrane (DPM) installed under the floor screed to enable moisture-sensitive floor finishes, such as wooden floors or laminates, to be laid on top.

Most modern DPMs consist of a two-part epoxy mix, which is spread evenly across the floor as a liquid using a notched trowel or roller.

You can also use a thick polypropylene SPM sheet, though this will raise the level of the floor considerably so may not be appropriate in domestic settings.

Ready-mixed screeds

Many manufacturers offer pre-mixed screeds that contain additional materials, such as combining a DPM or to retard or speed up setting. Some 'green' varieties replace ordinary Portland cement with recycled materials. The manufacturers produce data sheets outlining their benefits and how to use them – always make sure you know exactly what the screed will do before you start to work with it.

Screed rails

If there is no curb or edging from which to work, a screed rail or trammel bar should be placed in the bedding material at the required level. This is a straight, sometimes L-shaped rod, usually 3 to 4.5 m long. Many different types are available to buy, and may be made from uPVC, concrete, aluminium or steel. Some types are left in place to act as an expansion joint after the screed has set, while others are removed, or have a removable top.

Screed rails are firmly bedded onto the sub-base or laying course so that their top surface is at the level required for the surface of the screed. The level is determined by the compactibility of the laying course material and the thickness of the final floor covering. Rails need to be secure so that they do not move, settle or sag when the screed board is dragged over them, as this would mean your final screeded bed is at the wrong level. However, you still need to ensure that the rails can be removed, if necessary, before the final floor covering is applied.

PRACTICAL TIP

Always make sure you follow the manufacturer's instructions, especially when using more than one admixture together.

PRACTICAL TIP

Some floor screeding specialists use steel bars or lengths of 50 mm × 75 mm timber. The important thing is that the 'rail' is completely straight.

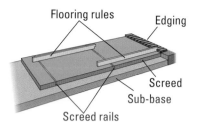

Figure 7.5 Positioning the screed rails

The screed rails should be aligned at 90° to the direction of the screed ruling, and spaced so that the screed board will span the gap between them with an overlap of around 200 mm.

Once they are in place, you need to set and check their level using a taut string line stretched between two known datum points, or with an automatic level.

Various beads and trims

We saw the different uses for beads and trims in Chapter 5. Similar items are used in floor screeds to form edges, define corners and embed movement joints.

Cement-based materials shrink as they dry, as well as expanding and contracting with changes in the temperature, especially where there is underfloor heating. To reduce the screed cracking, movement joints are often specified, especially if the final floor covering will be rigid, such as ceramic tiles.

Calculating areas, volumes and ratios of floor screed materials

Take another look at Chapter 2 to refresh your memory on how to work out areas and volumes.

To calculate approximately how much material you will need, you need to know the:

* specified surface area of the screed

* specified depth of the screed

* mix ratio of the screed, e.g. 1:3 or 1:4 (see 'Materials and resources used to form sand and cement screeds', below)

* density of the materials used (as a rule of thumb, both sand and ordinary Portland cement are around 1.5 tonnes per cubic metre but this varies according to the brand).

If you want to use fast-drying screed, you may also need to know the drying time.

PRACTICAL TIP

Typing 'ready mixed screed calculator' into a search engine will bring up various websites that will calculate the quantity of materials you need, based on the information you type in. The calculation will be based on an accurate figure for density of the materials used in that particular brand. However, you should only use the result as a rough guide as factors such as the moisture content of the screed and the type of sub-base will affect how much you actually use.

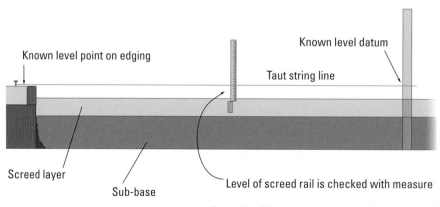

Figure 7.6 Checking the level of the screed rail

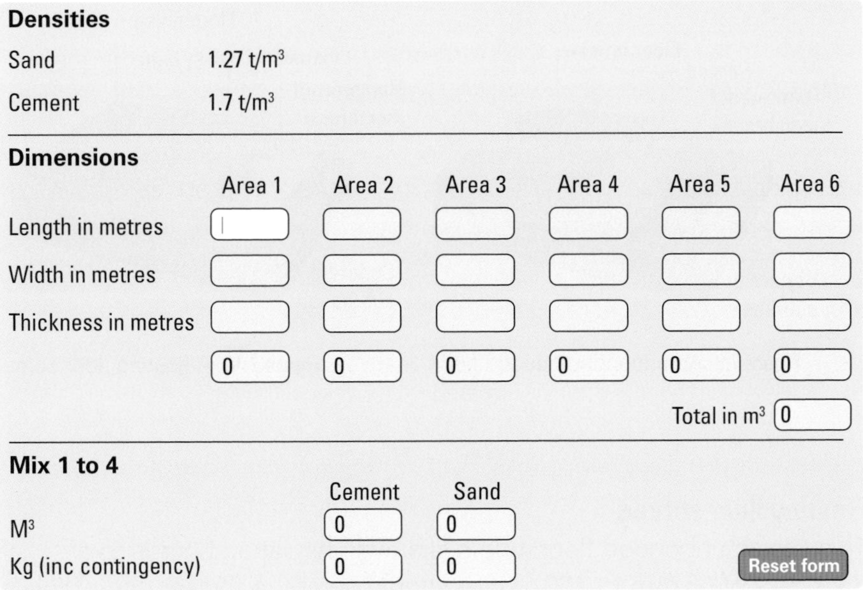

Densities

Sand $1.27 \, t/m^3$

Cement $1.7 \, t/m^3$

Dimensions

	Area 1	Area 2	Area 3	Area 4	Area 5	Area 6
Length in metres						
Width in metres						
Thickness in metres						
	0	0	0	0	0	0

Total in m^3 0

Mix 1 to 4

	Cement	Sand
M^3	0	0
Kg (inc contingency)	0	0

Reset form

Figure 7.7 An example of an online materials calculator (*Source* www.source4me.co.uk)

DID YOU KNOW?

BS 8204-2 provides guidance for installing and spacing joints in floor screed.

Types of floor screed

There are four main types of floor screed:

* bonded/monolithic
* unbonded
* floating
* separate.

PRACTICAL TIP

The better the preparation, the sounder the floor.

Bonded

This is the most common type. A hardened concrete base is prepared using cement and PVC, and ruled but left rough to help the screed bond.

The floor screed is then laid on top of it. The floor should be laid to a minimum thickness of 25 mm and would preferably be 40 mm thick. If it is too thick, however, the floor will not bond properly.

Fully bonded screeds are less likely to crack or curl so are the best type to use where there is going to be heavy duty usage or rigid flooring.

Monolithic

This is a type of bonded floor screed that is placed within 3 hours of the concrete base being laid. This will create a particularly strong bond with minimal shrinking because the concrete and screed will dry slowly together. It is between 10 mm and 15 mm thick.

Unbonded

Here, the floor is not bonded directly to its concrete sub-base, perhaps because of the concrete's poor condition or because there is a damp-proof sheet membrane laid on the concrete before the screed is applied. Its minimum thickness is 50 mm. This sort of floor is suitable for underfloor heating (if laid to 65 mm) or anywhere that might be affected by damp rising through the floor.

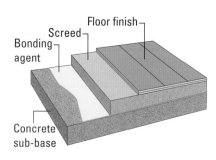

Figure 7.8 A bonded floor screed

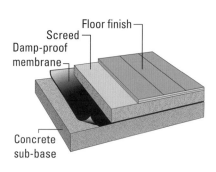

Figure 7.9 An unbonded bonded floor screed

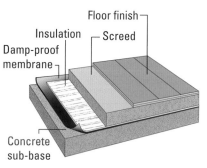

Figure 7.10 A floating floor screed

Floating floor screed

This like an unbonded floor, but is laid over insulation boards to a thickness of 65 mm to 100 mm. It should be at the thicker end of the scale in commercial locations, such as shops and offices, but can be thinner in domestic locations like houses.

Separate

This is a normal concrete floor where the screed has been left to bond to the surface. It is 40 mm or more thick.

PREPARING MATERIALS FOR FORMING SAND AND CEMENT SCREEDS

Laying floor screeds to levels and falls

Setting out levels

Whether the floor will be laid level or to a **fall**, the first task you must do is to strike a datum line around the area. Usually floors are laid level but sometimes they must be laid to a fall. In Chapter 2, we looked at using datum points as a reference for creating a level area. On a construction site, this should already have been worked out and marked. If this has not happened, or if you are at a private house, you will need to work out your own datum level, for example the bottom of a door. This is marked and then transferred around the building to wherever it is needed.

Whether the floor will be laid level or to a fall, you first need to mark the datum points at regular intervals around the room and then join them up using a chalk line.

Using a laser level

This is the quickest and easiest method. The laser beam is transmitted across the room as a red dot and you simply mark where it is before joining up your marks.

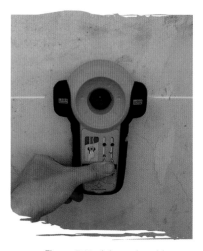

Figure 7.11 A laser level in use

Using an automatic liquid level

One way to form the datum line is by using an automatic liquid level. You must put the container of water at a height that is compatible with working areas for setting levels and falls; neither so low that it will get in the way of the floor, nor so high that you can't use it accurately.

Set the gauge on the rod to zero at the liquid level of the container, then move the rod to a point where you want to make a datum mark. Move the rod up or down until the liquid level is at the zero point you set on the gauge, then transfer this level by making a pencil mark on the wall. Repeat around the room then join up all the marks with a chalk line.

PRACTICAL TIP

If possible, set it on the floor, on a stool, on a box, on a pair of steps or on a high cupboard, such as a wall unit.

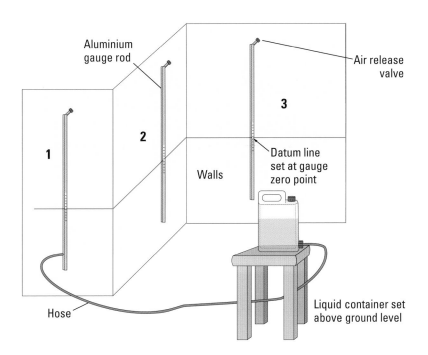

Figure 7.12 Using an automatic liquid level to mark datum points

Using a water level

This works on the same principle as the automatic liquid level but is simpler. Fix one end so that the water is at the height of the datum point. Put the other end where you want the second datum mark and wait until the water has settled. It finds its own level so will be at the same height as the first end. Mark this height on the wall and repeat around the room.

Using dots to level the screed

In Chapter 5 we saw how to screed a wall using dots. The same principle can be used for creating a level floor. Start by applying about half the required thickness of screed to a width of about 300mm. Compact it using the float to make sure all the material is pressed together and firm enough to receive a dot. It is best to make this dot fairly long. Then lay a second layer of screed and use a flooring rule to rule it in. Place a small timber batten or screed rail on it to the height

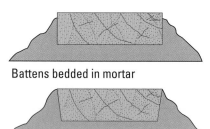

Battens bedded in mortar

Battens with sloping sides
are easier to remove

Figure 7.13 Batten placement in dots

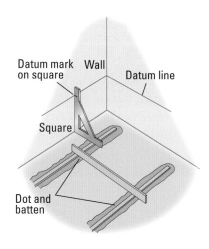

Figure 7.14 Levelling screed floors
using dots and battens

you need. If a thick floor covering is going to be used over the screed, such as tiles or floorboards, place one of these on top of the batten to get the final floor height. Use a builder's square to check the levels of the screed against the datum line or door frame.

Once you are satisfied that the first dot is accurate, set more dots at intervals around the room. When you set each one, check it is level by placing a long piece of timber between the dot and the previous dot and putting a spirit level on the timber. When you have placed dots all around the room, check again that they are level to within 3 mm over a 3 m distance.

Preparing floors with a fall

The dots method is especially useful for preparing floors with a fall, as you can set the dots at different heights. Set one dot onto the lowest part of the floor and then fix the rest of the dots to the line to give it a regular decline.

Another way is to use your flooring rule to help you. For example, if you need to lay a fall of 1:100, this is the same as 10 mm in every metre. So if the flooring rule or straight edge is 3 m long, place a dot or batten at 30 mm below one end of it. When you level the straight edge using a spirit level, this will mean that from one dot to another the floor will rise by 30 mm.

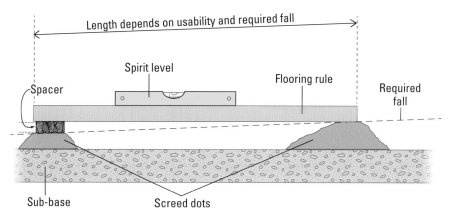

Figure 7.15 Using screed dots to achieve a fall

Gauging and mixing materials to the correct consistency

Most floor screeding contractors will tell you that getting the mix right is the most important element of the job. If it is not mixed correctly, the laying and finish of the screed will not be of good enough quality and the floor will be weak.

Gauging screed materials

The ratio of sand to cement is usually 3:1 – that is, 3 parts sand to 1 part cement. Sometimes 1:4 or 1:4.5 may be specified, and this is often the proportion used in ready-mixed screeds. Check this before you start to mix.

As we saw in Chapter 4, the easiest way to gauge the correct amounts is to fill a bucket with sand or cement, level it off and mix on the floor or in the mixer. To get a ratio of 3:1, you'd need 3 buckets of sand and 1 of cement.

Mixing screed materials

Floor screed material should be drier and more powdery than wall plasters and renders. However, you don't want it too dry as this will prevent a good finish. You also don't want it to be too wet, as this will affect the time the screed takes to dry and cause the cement to rise to the surface, which will weaken the floor. A wet mix will also create poor workability. The trick is therefore to add the correct amount of water.

Whether you are mixing by hand or in the mixer, add a little water at a time until the mix is damp. You can always add more water if you need to, but if it's too wet you'll need to add more materials, which is expensive and could affect the proportions.

If you are mixing by hand, make a well in the middle of your pile of materials and pour water into it. Then bring in the materials bit by bit with your shovel and mix them with the water.

If you using a mixer, put in a small amount of water first then add some sand and finally some cement. Repeat, adjusting the amounts as necessary until you have the correct consistency.

Table 7.4 describes some of the effects of not gauging and mixing correctly.

Problem	Possible effect
Not enough cement	The floor will be too weak so may crack or not be able to take a load.
Not enough sand	The mix will not bind properly, so will shrink, crack and fall apart.
Mix too dry	The screed is harder to lay. The finish will be poor. The floor will be weak.
Mix too wet	The screed is harder to lay. The screed takes longer to dry. The finish will be poor. The floor will shrink too much as it dries, and cement may rise to the surface, making it weak.
Insufficient mixing	The screed will be lumpy, preventing a flat, level surface and giving a poor finish. Clumps of sand prevent the mix from binding. Clumps of cement make the mix sticky and inconsistent.

Table 7.4 Effects of incorrect gauging and mixing

PRACTICAL TIP

Check your screed consistency with the 'snowball test'. Take a handful of the screed mix in your gloved hand and squeeze it. The screed should hold together without crumbling or dripping and should not stick to your hand.

Figure 7.16 The snowball test

PRACTICAL TIP

If a cement pump is going to be used, first check on the manufacturer's instructions that the mix can be used with a pump, and that you know what consistency is needed. The mix often needs to be wetter so that it can travel through the hose.

Protecting the work and its surrounding area from damage

Figure 7.17 Damaged floor screed

Newly laid screeds should be protected from impact and friction as damage to the screed could affect the quality of the final flooring. Remember that screeds are not the finished floor so are not intended to be used as a permanent surface.

The British Standards state that screeds should be protected from direct traffic (such as people walking over them, tools being placed on them and plant driving over them) as soon as possible after installation. Measures could involve clear signage, redirecting site traffic on large construction sites and installing screed protectors.

Protective measures should only be removed when the final floor covering is about to be installed. It is important to pay attention to the weight loaded on the screed and the level of compression of any insulation installed under the screed.

The time taken before the screed can take foot traffic depends on the brand and type of screed used. Generally it is between 2 and 48 hours – the exact time should be stated on the packaging or manufacturer's data sheet. Care still needs to be taken as the screed might not completely dry for weeks.

PRACTICAL TIP

Screed protectors are permeable plastic or polypropylene mats that enable moisture to escape from the drying screed. They are flexible and lightweight, so won't themselves damage the screed. Some are installed above the screed so can be walked on.

DID YOU KNOW?

Failing to protect a screed could not only damage it but also void any warranty on it. This should be made clear to clients.

DID YOU KNOW?

On some large sites, mobile elevating working platforms (MEWPs) travelling over the curing screed have squashed the insulation, which has ruined the screed. This can be avoided by checking all the specifications and other information at the start of the job.

LAYING SAND AND CEMENT SCREEDS

PRACTICAL TIP

Remember to start at the far end and work towards the door so that you don't get trapped in a corner and have to damage the screed by walking over it.

Laying screeds to levels and falls

Once you have prepared the sub-base, marked the datum levels and placed your dots, you are ready to lay the screed between the dots.

First brush grout onto the area you want to screed. The screed must be applied while the grout is still wet.

Figure 7.18 Ruling off the screed

Whether the screed is poured, shovelled or pumped, you need enough to spread over a large enough area to be able to rule it off and compact it. This means spreading the screed to a thickness of about 10 mm higher than the levelled edges.

Spread the screed with a flooring trowel or a shovel (for larger areas) and lightly compact it with a trowel or float.

Rule it off by pushing and pulling a flooring rule over the surface.

Try to keep the pressure even, so that the width of the floor is levelled equally.

Infill any holes with screed and repeat until you are satisfied that the section is ruled level and flat. Remove the dots or screed rails if they show above the screed.

You can then finish the floor with a float or trowel (see page 213).

Using bays

A helpful system to avoid getting stuck or having to walk on the newly laid screed is to use alternate bays in what is known as a chequerboard pattern. This requires separating the floor into squares or rectangles (bays) with a width to length ratio of 1:1.5, and laying alternate bays so that you can walk between them. Once the first set of bays is dry enough to walk and kneel on, you can come back to lay the rest of the bays.

The bays are made by dividing the room into sections using long timber battens embedded in screed. You could use your dots for this if you have laid them in an equal pattern. Fill in the spaces using a trowel and rule it off by resting your flooring rule across the battens.

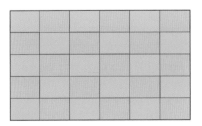

 Session 1

 Session 2

Figure 7.19 Bays for screeding

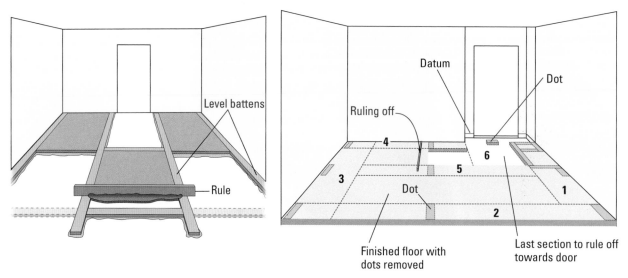

Figure 7.20 Screeding alternate bays

Figure 7.21 Rule and finish in the order shown

When you have filled in all the bays, remove the dots and battens then level them in, starting with the outside areas first. Remember to apply the finish as you work.

Laying floor to falls

Once you have established the levels, you can lay the falls. This might be a single slope downwards or all sides sloping towards a central drain or gulley. This drain should be fixed before you start to screed and, if you are using bays, these will be wedge-shaped if the falls are towards the centre.

Using self-levelling screeds

Self-levelling screeds are used to quickly and easily level or finish a floor that has minor cracks, lumps and hollows.

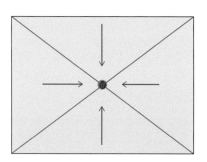

Figure 7.22 Work towards the centre of the fall

They can also be used as a finishing agent to give a smooth finish over traditional screed. Self-levelling screed is not a replacement for traditional screed as the overall quality of the finish is not generally regarded as being as good. However, it has several advantages:

* It is less labour-intensive.

* It can be laid to a shallower depth (generally 3 mm but cement-based self-levellers can be up to 40 mm).

* One bag covers up to 10 times more area.

* Shrinkage and curling are less likely to occur.

* Abrasion and impact resistance are better than for sand and cement floor screeds.

* It is less porous than concrete so will help with the adhesion of the finish.

There are disadvantages too:

* Self-levelling screed is not suitable where there are significant differences in height, such as large holes and bumps.

* You must work fast, as the mix dries very quickly.

* Despite its name, it takes some experience to get a level and smooth finish.

* It is very expensive, so you would need to balance the time saved against the cost of the materials.

Latex-based self-levellers are most commonly used, although you can also use water-based, acrylic or fibre-bonded types, depending on your sub-base. Self-levelling screed generally comes as a ready-mix to which you just add water. As with other types, the success of the final screed depends on it being mixed correctly.

It should be mixed more thinly than other types of floor screed, so that it flows across the floor. Add water and mix using an electric whisk until it's the consistency of tinned cream of tomato soup – try not to introduce air bubbles as these will remain when it dries.

Pour a small amount of the self-levelling compound onto the floor, starting with the lowest points, and use a trowel or spiked roller to join up the gaps. You don't need to get a perfect finish – as long as you have covered the area before the screed dries, it should set to create a reasonably smooth surface on which to lay the floor finish.

PRACTICAL TIP

You may need to apply a DPM or prime the floor first with PVA.

Compacting and finishing screeds

Compacting
Compacting the screed helps it to bond and stops it from falling apart. For smaller areas or in bays, you would do this with a float or trowel, but you can use a roller for larger areas. You need to apply the finish straight after you have compacted the area, before the screed starts to dry, to provide the appropriate key for the final floor covering.

Finishes
Floors are generally finished with either a floated or trowelled surface – whichever is suitable for the floor covering that is going to be laid. A wood float produces a sand-faced texture while trowelling gives a smooth finish which closes in the surface of a wood float finish.

PRACTICAL TIP

You can also use power floats, tampers and rollers.

Float finish
A large float is used to finish the surface if it is going to be covered with a heavy finish such as floorboards, mastic asphalt, asphalt tiles, bitumen or concrete tiles.

Trowel finish
A flooring trowel is suitable where a smooth surface is important, for example when the floor will be covered with carpet, hardboard, vinyl or cork tiles.

Figure 7.23 Float finish

Figure 7.24 Trowel finish

You can also use a latex self-levelling compound to give a smooth surface, as described above.

Curing screeds

If screed dries out too quickly, it is likely to curl and crack. The process can be slowed down by laying polythene sheets, building paper or damp hessian sacks over the screeded area to retain moisture during the setting process. The sheets should be left on the surface for a curing time of five to seven days before being removed to provide airflow.

Another option is to spray a chemical curing agent over the floor, which breaks down over a week or so as it assists the cure.

After this curing period, the floor should be allowed to dry naturally. Even if it looks dry, it could take anything from a few hours to 45 days for a newly laid screed to be dry enough to take the final floor covering (or even over 100 days for traditional screeds that have been laid thickly). Installing the floor covering too soon could spoil the whole floor, so it is best to conduct a moisture test using a moisture meter to ensure the screed has dried.

If your screed needs to be dry more quickly, for example if schedules are tight, use a self-levelling variety and consider running a dehumidifier to remove moisture from the screed. However, trying to speed up the process may not result in an adequate quality of finish.

Setting up drainage outlets in screeds

Wet rooms and showers may require you to incorporate a drainage outlet into the screed. You must lay any waterproofing and the screed to the correct fall to ensure that it is level with the outlet. If the floor is going to be tiled, make an allowance for the extra height by finishing the screed below the outlet or grate.

Remember that floor with embedded drainage outlets or gulleys usually need to be laid to a fall. Consult the specifications for details about the angle of slope that is required.

DID YOU KNOW?

One rule of thumb is that it takes 1 day for every 1 mm thickness of screed to dry, and 2 days per millimetre beyond a thickness of about 40 mm. However, the actual time will depend on the type of screed used and the temperature and humidity of the area where it has been laid. Fast-drying super-plasticisers dry at a rate of about 3 mm per day.

CASE STUDY

South
Tyneside *Homes*

South Tyneside Council's
Housing Company

You need practical skills as well as theory

Billy Halliday is a team leader at South Tyneside Homes.

'I didn't get great qualifications at school and I didn't have to take a numeracy or literacy test when I started out 30 years ago. I was just told to get on with it! Now there's much more theory and you do need the basic numeracy and literacy skills – using a tape measure, filling in forms, reading toolbox talks and putting them into practice – but the most important thing is an enthusiasm for the trade and mastering the practical skills. It comes down to being practically minded, being able to get your head down and work quickly and accurately on site.

I'd say to people thinking about being an apprentice, don't feel you can't do it if you don't have an A in maths. If you've got the right attitude, you'll learn on the job.'

BE ABLE TO PREPARE FOR FORMING SAND AND CEMENT SCREEDS

PRACTICAL TASK

1. PREPARING THE SUB-BASE FOR A BONDED FLOOR SCREED

TOOLS AND EQUIPMENT

Brushes	Flooring trowel
Buckets	Scraper

PPE

Ensure you select PPE appropriate to the job and site conditions where you are working. Refer to the PPE section of Chapter 1.

STEP 1 Ensure the surface of the concrete sub-base is clean and free from dust, rubble, loose plaster and anything else that would cause the screed not to bond properly. Scrape and sweep away loose dirt and dispose of it according to your site rules.

Figure 7.25 Scraping the floor

STEP 2 Brush the concrete sub-base with enough clean water to cover it. Leave the water to soak in, preferably overnight.

STEP 3 Brush off any excess water.

STEP 4 Mix the grout at a ratio of 1 part sand to 1 part cement with enough water to make a thick but workable consistency.

Figure 7.26 Applying grout to the concrete sub-base

STEP 5 Apply the grout to the damp concrete sub-base using a trowel, then brush it over to ensure a consistent covering.

PRACTICAL TIP

Smooth surfaces may need to be roughened to form a mechanical key.

STEP 6 When the grout has set, add a thin layer of PVA over it.

STEP 7 The sub-base is now ready for the floor screed.

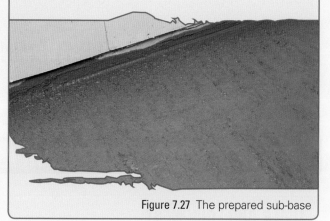

Figure 7.27 The prepared sub-base

PRACTICAL TASK

2. PREPARING THE SUB-BASE FOR AN UNBONDED FLOOR SCREED OR FLOATING FLOOR

TOOLS AND EQUIPMENT

Brushes   Flooring trowel

Buckets

PPE

Ensure you select PPE appropriate to the job and site conditions where you are working. Refer to the PPE section of Chapter 1.

STEP 1 Ensure the surface of the concrete sub-base is clean and free from dust, rubble, loose plaster and anything else that would cause the screed not to bond properly. Sweep away loose dirt and dispose of it according to your site rules.

STEP 2 Ensure the sub-base is flat, smooth and free from cracks or hollows. If necessary, repair these using small amounts of grout or floor screed.

STEP 3 The sub-base is now ready for the damp-proof membrane or insulation boards and floor screed.

PRACTICAL TASK

3. FORM A FLOOR SCREED TO LEVEL

OBJECTIVE

To practise the skills needed to lay a screed to levels, and apply a float finish to the surface.

INTRODUCTION

In this practical task, you will understand the types of floor screeds and materials used, and the quality of the finished surface. You will also understand the importance of datum levels and the need for accuracy in setting out.

TOOLS AND EQUIPMENT

Builder's square

Float

Flooring trowel

Shovel

Small piece of timber for dots

Spirit level

Straight edge

Water or laser level

PPE

Ensure you select PPE appropriate to the job and site where you are working. Refer to the PPE section of Chapter 1.

STEP 1 First check the floor is ready to receive the screed. Remove any loose particles and high points, and brush the whole area to remove the dust.

STEP 2 The next task is to strike a datum line around the area with a water level or a laser level. When you have levelled the lines on the wall around the room, join them together to form a datum line.

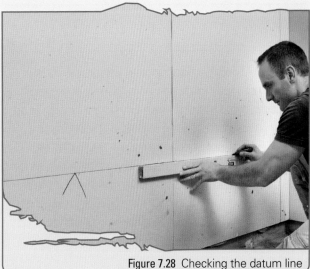

Figure 7.28 Checking the datum line

STEP 3 Now place the dots at 40mm height from the floor, working from the datum line. Compact it using the float so that when you place a small piece of timber on the screed, the timber and screed together are the required height of the floor. Then lay a second layer and use a flooring rule to rule it in. Place a small timber batten or screed rail on it to the height you need. Check the position of the dot with a straight edge and a level.

STEP 4 Start by applying about half the required thickness of screed to a width of about 300mm. It is best to make this dot fairly long.

STEP 5 Mix the floor screed according to the manufacturer's instructions or the specification if you are making it yourself. Using a shovel, add the material between the dots and compact it using a float until the screed is level with the dots. Now place the rule on the dots and, using them as a guide, rule towards you to fill in any hollows and rule it until the screed is straight and level with no uneven areas or hollows.

Figure 7.29 Ruling in the screeds

STEP 6 Once the screeds are completed, carry out the same process by filling in the main part of the floor area, using the screeds as a guide to rule off. Start to rub up the floor area with a float as you rule in, leaving a sand-faced texture, then trowel in the area to leave a smooth finish.

STEP 7 As you work towards the door, remove the dots and fill in any hollows.

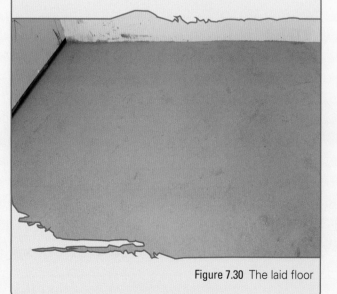

Figure 7.30 The laid floor

PRACTICAL TASK

4. FORM A FLOOR SCREED TO A FALL

OBJECTIVE

To practise the skills necessary to lay a screed to a fall, and to apply a smooth trowel finish.

INTRODUCTION

This method is used in wet areas like bathrooms and shower rooms, as you can set the dots at different heights.

TOOLS AND EQUIPMENT

Builder's square	Spirit level
Float	Straight edge
Flooring trowel	Timber for dots
Shovel	Water or laser level

PPE

Ensure you select PPE appropriate to the job and site conditions where you are working. Refer to the PPE section of Chapter 1.

STEP 1 Follow steps 1 and 2 from Practical Task 3.

STEP 2 Set one dot onto the lowest part of the floor using the method described in step 3 of Practical Task 3.

STEP 3 Fix the rest of the dots along in a line to give the floor a regular decline, depending on the fall required.

PRACTICAL TIP

For example, if you need to lay a fall of 1:100, this is the same as 10 mm in every metre. So if the flooring rule or straight edge is 3 m long, place a dot or batten at 30 mm below one end of it. When you level the straight edge using a spirit level, this will mean that the height from one dot to another will rise by 30 mm.

STEP 4 Place the drain or grid in position on the floor and ensure it is compacted into position.

STEP 5 Follow steps 4 and 5 in Practical Task 3, filling in the screeds to the correct height and ruling the floor towards the door area.

PRACTICAL TIP

Make sure you compact the material with a shovel and float, and apply a trowel finish.

STEP 6 Complete the floor with a float and smooth trowel finish.

TEST YOURSELF

1. What type of floor is laid on insulation boards?

   a. Monolithic floor

   b. Bonded floor

   c. Floating floor

   d. Separate floor

2. What sort of finish would you use if the floor was going to be covered with concrete tiles?

   a. Trowel finish

   b. Self-levelling

   c. No finish is required

   d. Float finish

3. What does DPM stand for?

   a. Damp-proof membrane

   b. Dots per metre

   c. Dust proof material

   d. Datum place mark

4. What sort of sand is best for use in floor screeds?

   a. Finely graded sand

   b. Sharp, well-graded sand

   c. Fine brick-laying sand

   d. Sea sand with shell particles

5. What is the classification of a level floor?

   a. + or – 30 mm in height over a 3 m distance

   b. + or – 1 mm in height over a 1.8 m distance

   c. + or – 3 mm in height over a 1 m distance

   d. + or – 3 mm in height over a 3 m distance

6. A screed rail must be...

   a. curved

   b. aluminium

   c. less than 200 mm long

   d. straight

7. What is the advantage of screeding in bays?

   a. You can walk between the bays so you don't have to stand on screed you have just laid

   b. It's quicker than laying screed over a large area

   c. Cement pumps can deliver the screed mix more accurately

   d. You can use a longer flooring rule to level the screed

8. When might a floor need to be laid to a fall?

   a. When the final floor covering will be wooden boards

   b. When stairs will be fixed to the floor

   c. When you are using a self-levelling screed

   d. When it requires a drainage gulley

9. Which of these is a disadvantage of self-levelling screed over traditional sand and cement screed?

   a. One bag doesn't cover as much area

   b. It is more likely to shrink

   c. It does not level out major defects in the sub-base

   d. It takes a long time to dry

10. How long should you leave a screed to cure under cover?

    a. About 100 days

    b. About a week

    c. 24 hours

    d. 45 days

Chapter 8

Unit CSA L2Occ56
PRODUCE FIBROUS PLASTER COMPONENTS

LEARNING OUTCOMES

LO1/2: Know how to and be able to prepare for producing plaster components

LO3/4: Know how to and be able to prepare materials for producing plaster components

LO5/6: Know how to and be able to produce fibrous plaster components

INTRODUCTION

The aims of this chapter are to:

* help you to understand the purpose of producing components from moulds

* show you how to create different types of mould

* explain the effects of different moulding materials and additives

* show you how to create components from different types of mould.

Fibrous plastering is a specialist area of the plastering trade and involves producing ornamental components from **moulds**. It also requires different tools from those you would use when working with solid plastering, and is a very different skill to master. Plasterers who specialise in producing moulded components are based in workshops and rarely, if ever, carry out other aspects of plastering, such as rendering or dry lining. However, it is worthwhile learning the basic skills, even if you do not go on to become a fibrous plasterer.

Components are **cast** from moulds off site, or at least away from where they are going to be located, before being fixed into position.

KEY TERMS

Mould

– a shape formed around an original object that is used to cast copies of that object or its parts. It may be made from metal, rubber, plaster or other materials.

Cast

– the reproduction or component obtained from a mould. May also be called a moulding.

Figure 8.1 A completed cast (left), made from a rubber mould (right)

PREPARING FOR PRODUCING PLASTER COMPONENTS

Hazards associated with producing plaster components

Casting plaster (plaster of Paris), as with standard plaster, is not dangerous if worked responsibly. However, plaster of Paris is classified by the Health and Safety Executive (HSE) as a hazardous substance so a risk assessment is required by law before it is handled.

Normally you will be casting using fairly small quantities of plaster. However, you still need to take care, because casting plasters give off a fine dust and can also irritate skin. Plaster of Paris is often used to take casts from body parts like arms or heads. Although it is unlikely that you will be asked to do this, do not assume that it is safe to use on bare skin, as it can heat up to 60 °C or more, and temperatures of just 45 °C can burn. It is a good idea to take the following precautions:

* Have a bucket of cold, clean water, a sponge and a towel available in case you need to rinse plaster splashes off your skin.

DID YOU KNOW?

The Health and Safety Executive risk management site and your own workplace will provide practical steps to follow when writing a risk assessment: *www.hse. gov.uk* has guidance and case studies to help you.

* When you have finished casting, clear up thoroughly and make sure no plaster dust remains that could cause breathing difficulties.

* Never pour wet plaster down the sink, even if it is only a small quantity. Ask your employer, client or site manager how to dispose of plaster materials.

* Place all plaster fragments in a rubbish bag and seek advice about how to dispose of it.

You may also use additives and chemicals in the moulding and casting process – many of these are toxic so, as with plaster, you will need to control their use, for example by using only the quantities you need.

Table 8.1 shows a sample risk assessment for working with casting plaster.

Hazard	Control measure
1. Plaster of Paris may irritate eyes, respiratory system and skin. It may burn when it heats up after mixing.	a. Wear gloves or a barrier cream when mixing and handling plaster of Paris. b. If handling large quantities of the dry powder, wear a dust mask to prevent prolonged inhalation. c. Wear eye protection to keep powder out of eyes. d. Do not be tempted to immerse any body parts in the plaster. e. Thoroughly wash hands and scrub nails after handling plaster of Paris. f. Ensure emergency eye-washing facilities are available, such as a clean hose attached to a cold tap.
2. Fragments of dried plaster, for example those loosened during cleaning moulds, can injure skin and eyes.	a. Wear eye protection when handling dried plaster. b. Cover up bare arms and skin. c. Clean area thoroughly as work progresses and after work has finished.
3. Mixed plaster can set and block sewerage systems if disposed of down a sink.	a. Plaster of Paris cannot be disposed of with other solid waste. It must be removed via a hazardous waste contractor or separated with other gypsum waste according to site rules.
4. Plaster of Paris and other chemicals used in fibrous plaster work could present a fire risk when stored.	a. Equipment and substances must be stored appropriately so as not to present a manual handling or trip, slip or fall hazard. b. Heavy items must be stored at the appropriate level. c. Ensure a COSSH risk assessment has been carried out if necessary. d. Flammable liquids should kept to a minimum and stored in a labelled, lockable metal storage bin or cupboard designed for the purpose.

Table 8.1 An example of a risk assessment for producing plaster components

Using information sources to produce plaster components

The two main things that you have to know before you start work are what the mould needs to look like, and how you should produce it. As always, refer to the layout, specification or block plan (see Chapter 2).

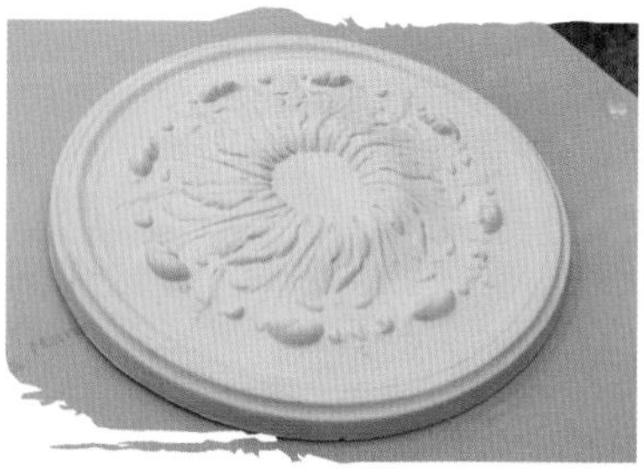

Using drawings and squeezes to produce moulded outlines

You may be producing the plaster component as a new feature, or because an existing item needs to be repaired or replaced. Table 8.2 shows the type of feature you may need to recreate. Note that you may know many of these as wooden features in modern houses, but they are likely to be made of plaster in historic houses. All these features may be fairly plain or highly decorative.

Figure 8.2 Ceiling rose

Figure 8.3 Cornice

Figure 8.4 Dado rail

Figure 8.5 Panel or plain-faced slab

Figure 8.6 Picture rail

Feature	Description
Architrave	The moulding around a window or door.
Beam case	A decorative moulding that covers a steel or wooden ceiling beam.
Ceiling rose (Fig 8.2)	A circular mould on a ceiling. The electric cables for a light often pass through it.
Cornice (Fig 8.3)	The uppermost part of a long, horizontal decorative moulding at the top of a wall, where it meets the ceiling.
Dado rail or chair rail (Fig 8.4)	The long horizontal moulding at chair height, originally to protect the wall from damage caused by chairs scraping against it.
Panel or plain-faced slab (Fig 8.5)	A flat, square or rectangular mould that is attached to the wall.
Picture rail (Fig 8.6)	The long horizontal moulding about two-thirds up a wall, from which pictures are traditionally hung.
Skirting (Fig 8.7)	The moulding at the base of the wall.

Table 8.2 Types of feature that may be moulded

There are three main ways to get enough information about the **profile** of the item you are going to copy or create:

1. By referring to a drawing – this may be full-size or drawn to scale.

2. By copying an existing piece, either directly or from a photograph.

3. By taking a **squeeze** of the item on site.

A squeeze is formed by pressing soft, wet, mouldable paper, pulp, latex or plaster or over the item you need to recreate. When the material is dry, it is removed and is a three-dimensional reverse (mirror) image of the original item. This may be used directly to produce the new item, or it may be converted into a drawing.

If it is a particularly decorative or complicated piece, especially if an area is being repaired, you could use all these ways of gathering information to ensure the match is as good as it can be.

Figure 8.7 Skirting

Types of plaster for producing fibrous components

As ordinary plaster is not fine enough to produce sufficient detail, you will need casting plaster. This is also known as plaster of Paris or Class A hemi-hydrate plaster. This has no retarder added and tends to set very quickly. Table 8.3 describes some different types.

Type of casting plaster	Description
Super fine	As its name suggests, this is extremely fine, so that you can produce very detailed patterns. It can also be used in combination with fine casting plaster.
Fine	This is the standard casting plaster, which is soft enough to be carved, sanded and shaped. British Gypsum fine casting plaster takes 18–22 minutes to set.
Coarse	This is usually used as a cheap way of coring out moulds (producing the inside part) but is not usually used where it can be seen because its finish is not as good as for fine or super fine plaster.
Autoclaved	This is used where a high strength moulding is required. It is made by making gypsum into a slurry and heated to around 220 °C to create a very hard plaster because it forms much longer and straighter crystals on setting. This is also known as an alpha plaster. It is made in smaller batches using specialist equipment so is expensive.

Table 8.3 Types of casting plaster

Each type of plaster has its own characteristics and it takes experience to achieve the best results. Brands to look out for are Teknicast, Helix, Herculite, Crystacal R and Crystacast. These two last types are extremely hard and dense so will produce durable casts that are not easily damaged. You will also need a hard grade plaster for reverse moulds so that they will not deteriorate when a large number of casts are taken from them.

If you need to maximise the strength of your cast, you can use architectural gypsum cement instead of standard casting plaster. This is often fire-resistant and is most suitable for large mouldings. Brand names include Hydrostone, Hydrocal and Ultracal.

Calculate quantities of moulding materials

The quantities of plaster and additives you will need depend on the type of plaster you are using and what the cast will be. It is also tricky to estimate the amount of water you will need for the mix. The amount of water used in a mix affects the strength, durability and setting time of the moulding.

Table 8.4 suggests ratios of dry powder to water for different grades of plaster but bear in mind that these are general recommendations and you need to check the manufacturer's instructions before beginning to mix.

KEY TERMS

Profile

– the shape of the item you need to copy.

Squeeze

– a method of obtaining a reverse profile.

DID YOU KNOW?

If you are using running moulds to repair or replace features like cornices in older buildings, you will probably use lime mortar to match the materials used in the original. Using gypsum-based mortar on lime-based mouldings will weaken both the original and new mouldings.

PRACTICAL TIP

You will find that many varieties of casting plasters are available from different manufacturers. For example, some are very hard and durable, containing glass resin, while others contain high levels of pure gypsum to give a very white finish for decorative plasterwork. Check the datasheet to ensure the plaster is suitable for your needs.

Type of plaster	Ratio of powder:water	Approximate setting time	Volume of water required per kg
Fine casting plaster	2:1.25	20 minutes	900 ml
Strong standard plaster e.g. Herculite No. 2	2:1	10 minutes	800 ml
Higher grade casting plaster e.g. Herculite stone	2.5:1	35 minutes	700 ml
Very hard-wearing plaster e.g. Crystacast	2.5:1	15 minutes	700 ml

Table 8.4 Suggested powder:water ratios when mixing moulding materials

In general, the finer the plaster, the more water you will need. Always check the manufacturer's data sheet and instructions on the packaging.

* **If you use too much water**, the mix will take longer to set and the plaster will be soft and crumbly.

* **Not enough water** may set the mix too quickly, and it could become full of air bubbles.

It can be difficult to get the proportion of plaster to water right across different batches, especially when you are new to fibrous plastering. However, it is important to ensure that batches are consistently strong and porous.

Gauging and mixing casting plasters

Gauging casting materials
As you are making individual mouldings, you will normally use a smaller amount of materials than you would when preparing a plaster for a wall or ceiling, so gauge boxes or buckets are likely to be too big for measuring out the materials you need.

However, it is more accurate to weigh the quantities of casting plaster rather than gauge them by volume, because bulk density can vary.

Mixing casting materials
As with standard plastering, it is important to mix the materials correctly to produce casts with the maximum strength. Your aim is to produce plaster with the consistency of single cream.

* Always use clean, drinkable water.

* Ensure your tools and containers are clean before you use them. Residue from dirty tools and containers may reduce the strength and working time of the plaster.

* Accurately weigh the plaster and measure the water to ensure uniform casts across different batches.

PRACTICAL TIP

Write down the powder:water ratio you use each time to get an idea of the best mix for that type of plaster.

PRACTICAL TIP

One rule of thumb for calculating how much plaster you need for a flood mould is to fill the mould with plaster then tip it into the mixing bowl.

PRACTICAL TIP

Remember to add the plaster to the water, not the other way round. Adding water to the dry materials will make the mix lumpy. Allow the plaster to absorb the water for a few minutes before adding more. This is known as soaking.

* Sprinkle in the powder rather than dumping it all into the water at once, to avoid lumps.

* For the best results, use a powered whisk of up to 400 rpm, set at a slight angle and not quite touching the bottom of the bucket or container.

* Once the mix has started to set, the material cannot be remixed so any leftover plaster must be disposed of.

Shorter mixing times will increase setting times but the plaster will be weaker and softer. Over-mixing can introduce air bubbles, which spoil the surface and reduce working time. In general, an ideal total mixing time is about 10 minutes and will prepare a batch of plaster with about five minutes of pouring time.

DID YOU KNOW?

Accelerators and retarders can be added to change the setting times and other properties of the plaster mix – see page 115. Note that additives may void the warrantee of the plaster product if a problem occurs.

Tools and equipment for producing plaster components

You need a variety of specialist tools and equipment to produce moulds. Table 8.5 shows the main equipment you will need.

Equipment	Description
Bench (Fig 8.8)	You need a bench big enough to run the mould you are making. Ideally it is about 3 m × 1 m or a metre square. It should be made of any strong timber and have a plaster or laminate (kitchen) top with a wooden or metal rule on each side. The top needs to be greased to stop the plaster sticking and to help the mould break free of the wood. You won't need your own bench when you're starting out, but you should check that the right type of benches are available on site.
Bowls (Fig 8.9)	It is useful to have some small bowls for mixing up small batches of plaster. These may be made from flexible, heavy duty rubber or polythene, so that you can easily press out dried plaster. Typically they have a diameter of 250 mm.
Buckets and tubs (Fig 8.10)	Use heavy-duty buckets not only to mix plaster but also to carry materials, water and waste. Useful sizes are between 25 and 65 litres.
French chalk (Fig 8.11)	A fine talc to prevent the mould from sticking. It will also stick to the grease inside the mould, so you can see if you have missed any areas. It is available as a powder and in sticks – powder is the best for moulding purposes but the sticks can be useful for marking the moulds or casts.
Plaster bin or box	Keep plaster tidy and dry by storing it in a special bin. As with a bench, you could make one yourself from plywood.
Scraper (Fig 8.12)	This is simply a flat, wide blade with a handle for scraping excess materials off the bench or out of mixing bowls. You can buy bent-bladed and extendable versions, but often an old trowel will do the job.
Water tanks (Fig 8.13)	You will need two tanks – one for mixing the plaster using clean water, and a slosh tank for cleaning your tools, moulds and equipment.

Table 8.5 Equipment for casting plasterwork

Figure 8.8 Bench

Figure 8.9 Bowls

Figure 8.10 Buckets and tubs

Figure 8.11 French chalk

Figure 8.12 Scraper

Figure 8.13 Water tank

As well as your standard tools (such as hammers, saws and small tools), you will need some specialist tools to help you make your moulding. Table 8.6 lists the main ones.

Tool	Description
Busk or drag (Fig 8.14)	This is mainly used to form and complete mitres (see page 260) and to shape and clean up mouldings. It is made from flexible steel and comes in various thickness and shapes.
Canvas knife	This is a sharp knife for cutting cut the canvas or hessian. You can also use scissors.
Gauging trowel (Fig 8.15)	This is used to mix small amounts of material and to get plaster into difficult places. It can also be used to clean down other tools. It is made of steel and usually has a wooden or plastic handle.
Joint rule (Fig 8.16)	This is to rule off plaster when making moulds and casts. Its working edge is the long, bevelled edge. One edge of the tool is cut at 45°. They come in a variety of sizes from 25 mm to 60 mm.
Running rule (Fig 8.17)	A long strip of timber used as a guide when making a running mould.
Sealant brush (Fig 8.18)	This is simply a brush for applying liquid sealant to the cast mould. Use a brush that is appropriate to the mould you are making, e.g. one with a small head will be better for getting into the crevices of a decorative mould, while a larger flat brush will coat large mouldings more quickly. Clean it carefully after use and don't use it for any other purpose.
Small tool (Fig 8.19)	Despite its general name, it is a tool in its own right. It is a flexible steel hand tool most commonly available in two designs: the leaf and square, and the trowel and square, and each design comes in three sizes of between 11 mm and 25 mm. They are used when small mixes are required and where there is small detailed work is to be completed. Trowel and squares are also excellent for stopping in (caulking) and leaf and squares are used for measuring out semi-viscous fluids like silicones and finishing off cast items.
Splash brush (Fig 8.20)	A long handled, round-headed brush used when casting from reverse mouldings. They are purpose-made for applying plaster onto reverse moulds. Make sure the brush is wet before it comes into contact with the plaster.

Table 8.6 Tools used for casting

Figure 8.14 Busk or drag

Figure 8.15 Gauging trowel

Figure 8.16 Joint rule

Figure 8.17 Running rule

Figure 8.18 Sealant brush

Figure 8.19 Small tool

Figure 8.20 Splash brush

PREPARING MATERIALS FOR PRODUCING PLASTER COMPONENTS

Materials used to form moulds

You will need to make several different types of mould for different purposes. A mould can either be:

* positive: the same shape as the item you want to cast

* reverse: a negative section of your cast, from which you make further casts to produce the item you need.

Running moulds

A running mould is a timber and zinc structure used to create a straight or circular panel or dado mouldings. The mould is the positive template, the plaster is poured onto the bench, and the template pulled or pushed over it to shape the plaster. Once it has set, the cast is fixed in place. You can learn more about running moulds on page 232.

Flexible moulds

These are usually made of a rubber-based material, which is poured over a profile to make a reverse mould. When the mould has set, plaster is poured into the flexible mould to create the positive mould.

When forming a flexible mould, you can use either **cold pour** moulding compound or hot melt compound (PVC).

KEY TERMS

Cold pour

–– flexible moulding compound that needs a catalyst to make it set, used in fibrous plastering.

Figure 8.21 Cold pour compound

Figure 8.22 Hot melting pot

Cold pour moulding compound

This is a flexible silicone rubber that, as its name suggests, does not need to be heated before use. It must be stored in a sealed container (a blast storage bin). This is because it consists of a chemical and a catalyst that you must mix together to begin the setting process.

You can buy different grades to suit the work you are doing. Table 8.7 shows the pros and cons of using this method.

Advantages	Disadvantages
The mould will be very strong and flexible	It cannot be reused so is expensive
It will provide accurate and detailed reproduction of the original model	It often takes experience to accurately measure quantities for the right mix
It is easy to prepare the cold pour on site	The chemicals used are toxic and flammable (may catch fire)
	Some cold pours may not set properly if they are contaminated with other things
	Both parts be thoroughly mixed to obtain a uniform cure/set

Table 8.7 Advantages and disadvantages of cold pour moulding compound

Hot melt compound (HMC)

Hot melt compound is a rubbery material known as polyvinyl chloride (PVC). It is a thermoplastic vinyl resin that comes in a variety of colour-coded grades depending on the flexibility you require.

To melt the compound, you need to use an electrical thermostatically controlled heating machine. PVC melts at 120–170°C, depending upon the grade chosen.

Table 8.8 shows the pros and cons of using this method.

Advantages	Disadvantages
Reproductions are of a reasonable quality	If the PVC is too hot, the model may release air bubbles that will affect the finish of the final cast.
The mould remains strong and flexible unless it is melted down too often	Hot melt machines are expensive, so the initial outlay may be more than small businesses or individuals can afford
You don't need to mix up the materials and measure quantities so less is wasted	Working with hot materials that produce fumes presents safety hazards
The mould can be cut up and reused or melted down when you have completed the job it was made for	

Table 8.8 Advantages and disadvantages of hot melt compound

Using additives

Numerous additives can be added to the basic plaster mix to change its properties, or used separately to make the moulding process easier.

Release agents

Release agents stop the cast from sticking to the mould, and they also act as a barrier between the casting media and the mould, to help prolong the life of the mould.

* Plasterers often keep a stock of grease to use as a release agent. Most casting shops mix it up themselves using tallow and paraffin or engine oil.

* A more expensive but less labour-intensive alternative is ready-made petroleum jelly (such as Vaseline) but these may inhibit the setting of glass fibre resin.

Retarders

Casting plasters tend to set in 15 to 20 minutes, which is often too quick to complete the cast. Retarders are used to lengthen the setting time. If you are making cornices, it means that you could mix up two plaster mixes at once and complete the job in one go.

* Traditionally, plasterers use glue size, which is a jelly-like material. The crystals are stirred bit by bit into very hot water until they have dissolved. Then hydrated lime is added to prevent it becoming solid.

* A less smelly alternative is trisodium citrate. Again, you can add a pinch of the crystals to water, or directly into the plaster mix if you are confident about how much you need. You don't need to add lime.

PRACTICAL TIP

Always use appropriate PPE, such as gloves, goggles and dust masks when using casting plaster and its additives. The chemical fumes in particular can be toxic, so it may be necessary to use a respirator or other barrier against breathing in the fumes.

Pigments

If you need to produce a coloured cast, you can use either powder or liquid pigments. However, you must be aware that pigments are likely to have other effects on the plaster. For example, some powdered pigments contain chemicals that accelerate or retard setting time, and introduce air bubbles. Only a small amount of pigment can significantly weaken the plaster.

PRACTICAL TIP

It is a good idea to test coloured pigments on a small batch first, not only to ensure you produce the right colour but also to find out whether the additive affects the plaster's set.

PRACTICAL TIP

You need to grease the whole area so make sure your bench is free from dust and old plaster before applying grease. Any small particles that get into the grease may scratch the mould and cause imperfections.

PRACTICAL TIP

Neither option has a recognised mixing ratio, as the level of retardation depends on the individual and the job. However, it's worth mixing up a test pot, as using too much glue size could slow the setting down considerably, while too much tri-sodium citrate will prevent the plaster from setting at all.

PRACTICAL TIP

As well as adding to the overall cost of the plasterwork, it is worth remembering that additives may affect the plaster in unexpected ways – changing its chemical make-up may prevent it setting or stain the plaster if not used properly. As always, check the manufacturer's data sheet before using them.

Add powdered pigments to the water before you add the plaster mix, and stir it to ensure the colour is consistently spread in the water. They will not necessarily dissolve.

Likewise, liquid colours are added to the water before the plaster is added. While it is less likely to weaken the plaster, it is harder to obtain a uniform colour throughout the finished cast.

PRODUCING FIBROUS PLASTER COMPONENTS

Creating reverse running moulds

To create a panel or cornice moulding, you must first create the running mould and then use the mould to shape the plaster. Running moulds are made up of two parts:

1. a template of the shaping being formed
2. a wooden frame (stock) for the template.

Making a running mould template

To get an accurate tracing of the moulding you want to copy, you either have to remove a section and trace round it, or make a saw cut across the profile, insert stiff paper or card into the cut, and draw around it.

Draw the profile you want to use on a piece of paper and glue it onto a plate of zinc. Using tin snips, carefully cut the zinc around the profile line and file it to make a clean edge. Check it against the drawing.

Lay the profile on the stock and draw around it with a pencil, then draw it again about 5 mm further out. Saw the wood to this second line to allow for swelling.

The running mould frame

The frame is constructed from plywood or timber with two straight edges. It is made of three parts:

* A stock – a support for the template, to prevent it from bending during the running.

* A horse or slipper – a support for the stock, which is fixed to it at a right angle (90°). The horse runs along the bench's running rule, and is 1.5 times as long as the stock.

* A brace – a support for the stock and horse to keep the frame from slipping. The plaster swells as it sets, so may try to push the frame out of alignment. Larger moulds may need two braces to minimise movement.

See Practical Tasks 1 and 2 on pages 242 and 245 for information on making a running mould frame and running a straight moulding from it.

See Practical Tasks 1 and 2 on pages 242 and 245

DID YOU KNOW?

All mouldings consist of members, which are a combination of curved and straight lines. Each member is separated by a straight section called a fillet.

PRACTICAL TIP

File marks will show on the final moulding so smooth these off using sandpaper or a metal nail. Zinc is very soft so you need to take care to get an accurate mould. Holding it still in a vice will help to prevent errors.

DID YOU KNOW?

For some very large projects, steel is used instead of zinc. As it is much stronger than zinc, it is sturdier, will last longer and will not bend. However, it is also much harder to cut.

It would normally be cast in situ on a wall or ceiling and may need coring out with heavy materials such as sand or cement.

PRACTICAL TIP

Splaying the outline's edge away from the template stops the plaster from clogging it up.

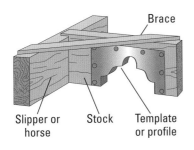

Figure 8.23 Example of a running mould of a simple panel moulding

(labels: Brace, Slipper or horse, Stock, Template or profile)

Figure 8.24 A circular running mould

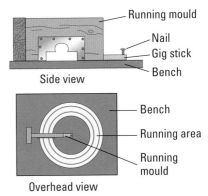

Side view

Overhead view

(labels: Running mould, Nail, Gig stick, Bench, Bench, Running area, Running mould)

Figure 8.25 A circular running mould using a gig stick

Running a circular mould

The method you will use to create a circular moulding depends on its size.

To run smaller circles, you will use an extended stock. The stock is long enough to extend from the slipper to beyond the turning point. A pivot point is fixed to the stock and the circle or part is run.

To run a large circular moulding, you also need a piece of timber known as a gig stick. This is attached to the stock and slipper, allowing the running mould to form larger circles on the bench. There are two ways of using this:

* The gig stick is slightly longer than the radius of the circle you need, and you nail it flat on top of the stock and hand brace. Then the stick is nailed to bench table at the other end so that it can go round in a circle.

* The gig stick needs to be long enough to extend from the slipper to beyond the turning point. Then, to allow the running mould to rotate, you need to fix a turning eye at the end of the gig stick. This is a piece of metal with a 50 mm hole in it. Then a turning block is nailed below the turning eye to attach the mould to the bench. It can be further held in place with a piece of canvas soaked in plaster (a wad).

The procedure for running these types of running moulds is the same as for any running mould except that, when running a complete circle, you must take care to remove all excess plaster on the final run.

> **PRACTICAL TIP**
>
> Take care with this method as the angle of the gig stick could cause it to foul (get in the way of) the running profile.

Figure 8.26 Using a circular running mould

Figure 8.27 Building up the circular profile

Figure 8.28 The finished circular moulding

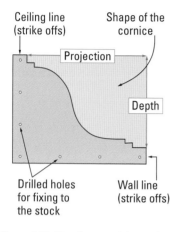

Figure 8.29 Cornice mould template

Labels on figure: Ceiling line (strike offs); Shape of the cornice; Projection; Depth; Drilled holes for fixing to the stock; Wall line (strike offs)

KEY TERMS

Muffle

– a false profile made of plaster, metal or wood, fixed to the metal profile on a running mould.

Forming a reverse cornice mould

If you need several identical items, you would use a reverse mould to create a cast from which you run out the mouldings you need. You form a reverse cornice mould more or less in the same way as you would a panel moulding. The profile would match that of the cornice you want to replicate.

As the moulding is bigger than a standard panel, the template needs to be larger, and this means that you need to add a **muffle** to maintain the shape.

The muffle is fixed to the metal profile on a running mould. The muffle is usually shaped from plaster, because it is easy to mould it to match the existing profile, but you can also use metal or plywood, fixed about 6 mm in front of the metal profile of the running mould.

If you are making a plaster muffle, first you need to fix nails to the stock, leaving them sticking out slightly.

Mix your casting plaster thicker and stiffer than usual, and start to apply it to the profile, gradually building it up to follow the shape of the profile. A small tool is useful for this.

When you have built it up to 5 or 6 mm above the profile, leave it to set for a few minutes, then shape it to match the profile.

Before running, apply shellac (see page 241) and grease to the muffle so that the cast does not stick to it.

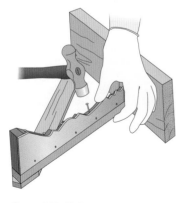

Figure 8.30 Fixing nails to the stock for the muffle

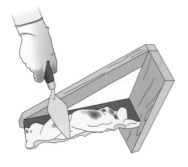

Figure 8.31 Building up the muffle

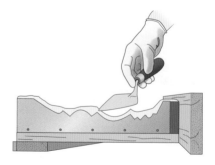

Figure 8.32 Shaping the muffle

As well as a muffle, you will need to make a **core** to support and bulk out the moulding.

You are likely to need to use a core when the moulding section is not thick enough to support the expansion of the plaster when it sets, which would stop the movement of the run. It is also useful when an undercut section of the moulding would be difficult to fill in one mix.

You can form the core using a variety of materials:

* strips of plasterboard
* old plaster mouldings
* laths soaked in water
* hessian canvas soaked in plaster
* casting plaster
* EML, sand/cement and lightweight plasters are often used for very large moulds.

You might also use a combination of these – for example, a plasterboard core with plaster-soaked hessian laid over it.

When you have poured the plaster over the core and run the running mould over it, you will form the shape of the muffle. You then take off the muffle, clean the running mould and, once the plaster has dried a little, key the core with a craft or trimming knife. Then re-grease the bench and run the mould again, this time against the true profile.

When the mould matches the profile, wait for it to dry, cut it to size and give it a few coats of shellac (see page 241). You should also grease it to act as a release agent.

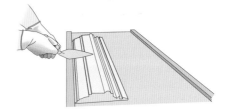

KEY TERMS

Core

– a build-up of material on a moulding to leave only a finishing thickness to a running profile used with a muffle.

PRACTICAL TIP

Grease the bench before forming the core, and nail it to the bench so it does not move during the running.

Figure 8.33 Forming the core

Figure 8.34 A model of a reverse cornice mould, showing the core and different stages of build up

Figure 8.35 Demonstrating the final fit of the profile

Loose piece moulds

Loose piece moulds consist of several separate parts, which can be added or removed as required. They are mainly used when making casts for fibrous restoration, or those that require intricate details or features that are not standard.

Using firstings and seconds

To run a mould using plaster, you can use either the two-gauge or the one-gauge (sometimes called single-gauge) casting system.

Two-gauge casting system

This is the system used for the majority of moulds, including those described in this chapter. It uses two plaster mixes, called firstings and seconds. As you might guess, the firstings are mixed and applied first, and the seconds are applied along with the canvas after the firstings have set. The firsts are slightly thicker, and the seconds are mixed with size or another retarder to slow the set. The idea is that a second layer gives a better finish because otherwise the canvas might show through on the face.

One-gauge casting system

This uses a single mix of plaster with size, which avoids two mixes swelling at different rates but has the disadvantage of risking pressing the canvas through to the face.

Materials for reinforcing plaster mouldings

Moulds made from plaster of Paris are fragile and so may need to be reinforced to ensure they last as long as they are intended to. If the mould is made in a workshop off site, it needs to reach its destination in one piece. Additional support might also be needed during the casting process, to ensure an accurate reproduction is made. Other than using cement-based plasters (as discussed above), plaster can be strengthened in a number of ways.

Canvas or hessian

This is used as a standard reinforcement for plaster casts. It can vary in size from 75mm up to 900mm and comprises of a 3 or 6mm square mesh jute weave of which can be cut using scissors to any length required.

Wads

Wads are canvas strips that have been soaked in plaster and stuck to the back of the cast, along its length. They may have wire embedded in them, which is fixed to the background. The slab can be levelled up or down by lengthening or shortening the wire.

Ropes

Ropes, like wads, are made from canvas cut into lengths and soaked in seconds, but then squeezed and twisted to form ropes. They are usually placed under laths to reinforce the plaster casts and form a strike off.

Figure 8.36 Hessian

KEY TERMS

Wads

– square pieces of scrim cloth soaked in casting plaster and used for making and fixing casts and reinforcing joints.

Laps

These are strips of canvas that go across the mould. They are especially useful in reinforcing a cornice across its curve. You can use canvas offcuts, folded and cut across the fold, to make several smaller squares of canvas. Fold these and soak them in plaster. They can be fixed to the back of the cornice about 300 mm apart.

Laths

Laths are strips of sawn timber to reinforce and fix the casts in the manufacture of plasterware like cornices, picture rails, dado rails and ceiling roses. When casting, they can also form the strike offs (wall and ceiling lines).

They are 3 mm or 5 mm thick, 22 mm or 32 mm wide and about 3 metres long, although many suppliers make them to order and supply them in bundles of 100 or so.

You need to soak your lath prior to use to prevent it twisting, cracking and **crazing** on the plasterware's decorative face. It is also worth adding a mild bleach solution to the lath tank to kill any mould spores on the wood.

Timber bearers

These form the frame that encloses the moulding, especially for a plain-faced cast. They will form the strike offs for the mould. They need to be a little larger than the size of the intended cast, and at least 300 mm × 300 mm.

Fibreglass

Chopped glass fibre strands can be added to the mix or, for larger or less detailed moulds, fibreglass mats can be incorporated into the cast.

Figure 8.37 Preparing canvas strips for reinforcing a mould

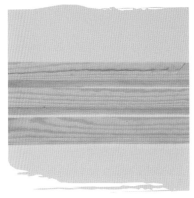

Figure 8.38 Laths

KEY TERMS

Crazing

– hair line cracks on newly plastered and rendered surfaces.

PRACTICAL TIP

The laths must be damp at all times when you apply plaster so you will also have to spray them as you work.

The effects of incorrectly positioned reinforcing materials

You must make sure canvas and timber reinforcements are cut to the correct size and positioned accurately. If not:

* the cast might be weakened and cracks appear

* they might show through the plaster

* they might extend beyond the edges of the mould

* it will be difficult to transport the cast to the site of the job for fixing.

It is worth thinking carefully about where the reinforcements will go, and checking they are in the right place, before applying seconds.

Figure 8.39 Fibreglass reinforcements

Casting plain plaster mouldings from mould templates

Casting a plain-faced slab

You may need to produce a comparatively large expanse of plain plaster moulding, for example to patch a ceiling. This is comparatively straightforward and practising the skills you need to produce a plain-faced slab will make more complex casting easier when you come to do it.

Timber bearers or rules form the frame that encloses the moulding, especially for a plain-faced cast. They will form the strike offs for the mould. Measure the area you need to cast on the bench – the frame needs to be a little larger than the size of the intended cast, and at least 300 mm × 300 mm.

When you have checked your frame with a builder's square, nail or screw it to the bench. These timber edges become the strike-off points.

Cut your backing and facing canvas, and any ropes, to length. You should also cut your pre-soaked laths now – you need four pieces, that fit on the inside edges of the mould, parallel to each side of the frame.

Mix the firstings and seconds, then pour the firstings into the centre of the fenced area, and let the plaster slow to the edges.

Use a joint rule to ensure the whole area is covered with an even layer of plaster.

When the firstings have set, add the reinforcements – the hessian and the laths.

Now pour the seconds to cover the reinforcements, and rule the plaster off so it is level with the timber edges (strike-off points). Clean off the strike-off points with water then splash a final layer of plaster over the cast.

When the slab is dry, remove the timber bearers and you will have a strong, flat slab of reinforced plaster.

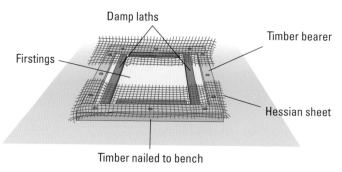

Figure 8.40 Adding the reinforcements to the slab

Casting a ceiling rose

You use a similar process to make a flat cast from a reverse mould, for example using a PVC flood mould to create a ceiling rose. You simply pour the **firstings** into the reverse mould and spread a layer about 3 mm with a splash brush. The main difference is in the preparation of the mould, which (if you are using PVC) would not require shellac (see page 241) and grease, but reinforcing materials are still required along with the same mixing procedure. You do need to grease moulds made of non-flexible material.

When the firstings have nearly set, add a layer of seconds and then any reinforcements, which you cover with more seconds and a final splashed layer.

The practical task on page 250 shows you how to cast flood moulds.

The practical task on page 250 shows you how to cast flood moulds.

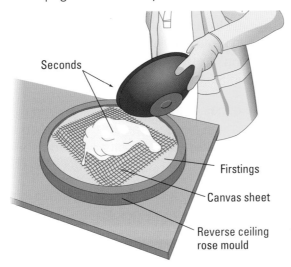

Figure 8.41 Adding seconds to a reverse flood mould

Casting using a reverse cornice mould

Once you have created your reverse cornice mould or plastered core (see page 248), you can use it to produce the casts you need. This is a fairly straightforward process, but its success, as is often the case in plastering, depends on good preparation.

Cut a piece of hessian to the length of the mould, plus about 100 mm at each end. It should be wide enough to fold it back into the centre of the core.

Ensure that you have enough soaked laths ready, and that they are the right length.

Make up the firstings and seconds. The firstings should be a little thicker than the seconds, and the seconds should contain a retarder (which makes it convenient for you to mix both at once).

Pour the plaster onto the mould and spread it with the splash brush. You can initially brush this on, but splashing it will produce better coverage and reduce air bubbles.

KEY TERMS

Firstings

– the first coat of casting plaster applied on to a cast.

PRACTICAL TIP

To get a straight line when cutting canvas, pull out two of the threads – this will leave a guide line.

PRACTICAL TIP

Make sure everything is in place before mixing the plaster as you won't have time to do it once the plaster begins to set.

Figure 8.42 Splashing on the firstings

PRACTICAL TIP

Fan the splash brush with your fingers when applying the plaster as it will cover a larger area.

When you are happy with the coverage, clean the plaster off the strike offs (wall and ceiling lines) – these are the edges of the cast.

PRACTICAL TIP

Run your fingers along the arrises to remove any air bubbles, which will make holes in the cast.

As the plaster begins to set, lay the hessian over the whole mould. You can also add laths to provide a flat, solid surface for fixing the cornice to the wall and ceiling. You might also need to add a curved lath parallel with the strike-off area.

Figure 8.43 Building up the layers

Now fold back the edges of the hessian over the mould, as you do not want it to stick out of the sides. Clean off the strike-off area again but don't remove all the plaster – you eventually need to build it up here to about 20 mm.

Pour over the seconds and rule the plaster off with a joint rule so that the plaster is level with the striking points.

Rule off the strike offs and leave the cast to set. Once it is cold and dry, carefully separate it from the core mould – this is a delicate procedure so you need to be careful. Slide your fingers along the edge and tilt it towards you. Wipe the grease off the mould. You can now clean off the core mould to use again.

REED TIP

Learning plastering skills is a lifelong experience; we're all still learning, and even after your apprenticeship is finished, you'll keep on learning too.

Sealing and seasoning models and moulds

You must make sure the plaster cast is completely dry before sealing it. You can either leave it to dry in the air or use a fan oven, set at 45–50 °C. This provides a more uniform and controlled drying environment than air-drying, but take care not to over-heat it as the water will evaporate too quickly, leaving the cast powdery and weak.

If you stand a mould upright, on the smallest surface area, air can get all around it, which prevents warping.

When the moulding is dry, you can either leave it as it is or seal it. Plaster is porous so, if you need to paint the moulding, sealing or priming it first will give you a better finish and protect it from water damage and mould. It is a good idea to apply a water-resistant seal if the moulding is going to be fixed somewhere that might be at risk of damp.

DID YOU KNOW?

Just because it feels dry on the outside doesn't mean it is dry at the core. Plaster generates heat as it dries, which causes surface water to evaporate, but water in the centre of the moulding will take longer to make its way out.

Shellac is most commonly used. It is a painter's polish or varnish that seals the plaster surface and makes it waterproof. You dilute with methylated spirit at a ratio of 1:3 or 1:4 and then apply it in three or four coats.

DID YOU KNOW?

Shellac is a resin made from the secretions of the female lac insect. It is scraped from the trees so the insects are not harmed. It takes about 100,000 lacs to make 500 g of shellac flakes.

PRACTICAL TIP

Sealing the moulding won't stop it from being damaged if you drop it. Always store and transport casts with care.

Although shellac is usually the preferred sealant, it is expensive, labour-intensive to apply and scratches easily, so, depending on the moulding's final use and position, you can seal it in several other ways.

* Spray the mould with an **acrylic sealer**.

* For a clear, slightly glossy finish, apply **epoxy resin** or **PVA** diluted with water.

* **Polyurethane** provides a tougher finish and is available in gloss, satin or matte.

* If the moulding will be painted, first apply a thin wash of **diluted paint** as you would when decorating a newly plastered wall.

* You can also brush on a proprietary solvent-borne sealant.

Sealing is the last stage of the **seasoning** (preparation) process.

KEY TERMS

Seasoning

– preparing of moulds.

Protecting your work and the finished casts

If you are on a large site with other trades, it is best to carry out moulding work in a separate workshop. If this is not possible, you will need to agree an appropriate space with the site supervisor, where people are not going to interrupt or accidentally damage the mouldings as they pass.

Completed casts should be stored in a cool, dry place, and covered if possible.

If they are not yet dry, consider where you will need to put them so that they are not squashed out of shape or broken.

As with plasterboard, transport casts on edge. They are very easily damaged so use two people to carry them if possible, and ensure that the people transporting them are aware of how fragile the casts can be.

Try to store them flat, although cornices should rest against a wall or scaffold board. You may need to the tie them in place to stop them sliding down.

Put away tools and moulds when you leave site, to prevent accidents, theft and damage. The store should be secure and watertight.

DID YOU KNOW?

Cornices should be stored upright, face to face for a pair, and with pairs back to back with another pair if you have four or more casts.

CASE STUDY

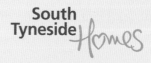

South
Tyneside

South Tyneside Council's
Housing Company

Further training enhances your job prospects

Glen Richardson is a final year apprentice at South Tyneside Homes.

'You learn new skills on the job every day because every job is different. Obviously you're learning new skills in college too, but if you're given the chance to do any further training with your employer, then that's something extra you can put on your CV. It will be transferrable to other jobs too.

I've done a manual handling course on how to pick up heavy and awkward objects correctly and how to store them without injuring myself. It's definitely improved the way I work.

Even more useful was the scaffolding qualification I got. I can now erect and dismantle scaffolding, which is something you're not automatically allowed to do without the proper training. This means that I get to work on bigger jobs and buildings, and it will definitely help if I ever need to look for another job.'

PRACTICAL TASK

1. CONSTRUCT A RUNNING MOULD

OBJECTIVE

To draw a moulding section from information provided using basic geometry, and then to construct a running mould by transferring the moulding outlines onto metal profiles using woodworking skills.

INTRODUCTION

Running moulds are made from timber and metal. They are used to form shapes in plaster. These plaster mouldings are run in neat plaster of Paris for maximum strength.

You can run straight or curved mouldings using different techniques.

TOOLS AND EQUIPMENT

Pin hammer	Zinc sheet
Coping saw	Tin snips
Tape measure	Nibblers
Trimming knife	Set square
Scissors	Files
Profile paper	Vice

PPE

Ensure you select PPE appropriate to the job and site conditions where you are working. Refer to the PPE section of Chapter 1.

STEP 1 Draw a basic profile onto paper, making it no more than 100 mm long and 50 mm high.

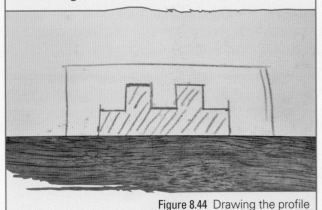

Figure 8.44 Drawing the profile

STEP 2 Place the paper profile over a piece of zinc and mark out the profile by drawing around it onto the zinc.

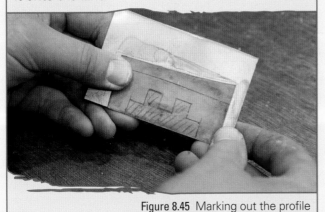

Figure 8.45 Marking out the profile

STEP 3 Use snips to cut out the basic shape and nibblers to cut out the profile on the zinc to approximately 2 mm outside the line.

Figure 8.46 Cutting the zinc with snips

STEP 4 Place the cut zinc in a vice and, using small square and round metal files, carefully file down to the line to form a clean cut zinc profile.

Figure 8.47 Filing the profile

PRACTICAL TIP
Using wire wool, clean off any edges on the zinc profile to leave a smooth, sharp, clean edge.

STEP 5 Saw the pieces of timber to length. For basic running moulds, the stock should be 100 mm high by 150 mm long. The horse/slipper should be one and half times the length of the stock (so it needs to be 225 mm if the stock is 150 mm long).

Figure 8.48 Cutting the timber for the stock

PRACTICAL TIP
Make sure you cut all timber square. Use sandpaper to give a smooth finish.

STEP 6 Now draw round the zinc profile onto the stock. Then draw a second line 5mm away from the profile line.

STEP 7 Place the stock in the vice and use a coping saw to cut out the timber to the second line.

Figure 8.49 Cutting out the profile shape

STEP 8 Now fix the zinc profile into position onto the stock using panel pins.

Figure 8.50 Fixing the zinc profile to the stock

STEP 9 Measure to the centre of the horse/slipper lengthwise and mark it. Drill a hole through the point you have marked and screw the stock and slipper together.

Figure 8.51 Fixing the slipper to the stock

STEP 10 Cut a piece of timber to form a brace. This needs to be long enough to stretch from the far top edge of the stock to one end edge of the slipper. Make sure the timbers are square then fix the brace to the slipper and stock using screws or nails.

Figure 8.52 Fixing the brace to the stock

STEP 11 Sand down the running mould and ensure it is square and flat. It is now ready to run.

Figure 8.53 The finished running mould

PRACTICAL TASK

2. RUN A STRAIGHT SECTION OF MOULDING

OBJECTIVE

To mix the casting plaster and to produce a small mould such as a dado or panel mould, using the running mould you made in Practical Task 1.

INTRODUCTION

The running mould is run against a running rule to ensure that the moulding is run straight. You must make sure that the moulding is run on the bench and prevent any materials gathering underneath the running mould.

TOOLS AND EQUIPMENT

Small tools	Mixing bowls
Drag or busk	Pin nails
Pin hammer	Casting plaster
Joint rule	Running rules
Splash brush	Grease

PPE

Ensure you select PPE appropriate to the job and site conditions where you are working. Refer to the PPE section of Chapter 1.

STEP 1 Form a running rule by fixing a straight piece of timber about 1.8 m long to the bench using screws or nails.

Figure 8.54 Fixing the running rule

STEP 2 Mark out the length of moulding required on the running rule.

Figure 8.55 Marking out the length of moulding

STEP 3 Check the bench area is clean and free from hollows. If it is not flat, fill in the hollows with a little plaster and leave it to set before starting to run the mould.

STEP 4 To keep the mould in place, fix anchorage nails using the lengths marked on the running rule as a guide, ensuring that they are positioned in the thickest part of the moulding section but don't put one at either end, as they might snap the plaster as it swells. Check that the moulding does not catch on the nails before preparing the mix.

PRACTICAL TIP

To make it easy to release the plaster from the nail once you have completed the run, put a little pyramid of clay or a lump of 'sticky tack' or putty over the head of the nail.

Figure 8.56 The anchorage nails and running rule being fixed in place

STEP 5 Evenly grease the whole area where the plaster will be applied.

PRACTICAL TIP

It is especially important to grease the running mould around the zinc profile area, as this will help the running mould to run freely on the bench.

Figure 8.57 Greasing the bench

Figure 8.58 Greasing the running mould

STEP 6 Mix a full bowl of plaster of Paris until it is the consistency of single cream. Use a small piece of lath or small tool to mix the plaster.

PRACTICAL TIP

Some plasterers prefer to mix the casting plaster with their hands. If you do this, make sure you are wearing latex gloves as the mixed plaster heats up, which could burn your skin.

PRACTICAL TIP

The amount of plaster you need will depend on the size of the section of moulding. It takes some experience to get it right. If in doubt, mix a smaller amount but try to ensure that the bulk of the moulding is completed with the first mix.

Figure 8.59 Casting plaster mixed to the correct consistency

STEP 7 Pour the mix from the bowl onto the bench area. Place the running mould at the far end with the template facing along the length of the running area. With one hand on the hand brace, firmly pull the mould against the running rule as you walk backwards, keeping it flat on the bench. Catch the excess plaster in a bowl at the end of the bench so that you can use it again. As members (straight and curved shapes) build up, add more plaster.

Figure 8.60 Pouring the mix

Figure 8.61 Running the mould

STEP 8 If another mix is required, follow steps 6 and 7 using a smaller amount of plaster. Make sure you wash the mould after each run and clean the edges of the mould with a joint rule. The mould is ready when it fits the profile of the running mould without any hollows.

Figure 8.62 Washing the mould

STEP 9 Once the moulding has fully set, cut it to length using a small-toothed saw and plenty of clean water to damp it down and reduce dust.

Figure 8.63 Cutting the moulding to length

STEP 10 Finally, lift the mould carefully from the bench using a joint rule and store it on a flat, dry surface, away from where it might be damaged. Ensure you remove the anchor nails from the bench.

Figure 8.64 The finished moulding

PRACTICAL TIP

The procedure for running circular running moulds is the same except, when running a complete circle, you must take care, when you have finished the final run, to remove all excess plaster.

3. PRODUCE A CORNICE REVERSE MOULD

OBJECTIVE

To prepare a running mould and produce a cornice reverse mould from it.

INTRODUCTION

Producing a cornice reverse mould reduces the weight of the plaster and allows you to reproduce any type of moulding that the client requires. All casts for a cornice reverse mould need to be reinforced with hessian scrim and may also need timber reinforcement. If this is the case, the timber laths need to be soaked in water overnight.

TOOLS AND EQUIPMENT

Small tools	Splash brush
Drag or busk	Mixing bowls
Scissors	Pin nails
Pin hammer	Casting plaster
Joint rule	Running rules

PPE

Ensure you select PPE appropriate to the job and site conditions where you are working. Refer to the PPE section of Chapter 1.

STEP 1 Apply a coat of shellac to the reverse mould and grease it well to allow the completed cast to be easily removed.

Figure 8.65 Shellac and grease on the reverse mould

STEP 2 Cut the hessian front and back canvases, allowing an overlap of 50 mm beyond each side of the mould.

STEP 3 Cut the pre-soaked timber laths to the length of the moulding.

STEP 4 Mix the seconds before the firstings, because you add retarder to them. Leave the seconds to soak to be used later.

STEP 5 Now mix the firstings to the consistency of double cream.

STEP 6 Pour out a small amount of the firstings onto the moulding and brush it over the whole area.

STEP 7 Now, holding the bristles of your splash brush, splash a layer of plaster up to 3 mm thick all over the reverse mould.

Figure 8.66 Splashing on the firstings

Figure 8.67 Remember to clean the strike-off points at every stage

Figure 8.70 Applying seconds over laths and canvas

STEP 8 Position the facing canvas and, using your splash brush, soak this canvas with a layer of seconds.

Figure 8.68 The facing canvas in position

Figure 8.71 Keep cleaning the strike offs

STEP 11 Splash a final layer of plaster over the cast to give it extra strength and thickness.

Figure 8.72 The completed cast

STEP 9 Now add the timber laths on each side of the moulding and splash with the firstings.

Figure 8.69 Covering the canvas in the seconds

STEP 10 Add the backing canvas and pour over the seconds, brushing them in and covering the entire surface area.

STEP 12 Let the cast set for approximately 30 minutes then remove the cast, carefully lifting it from the mould. Clean off any excess plaster with a joint rule.

Figure 8.73 The completed cornice, lifted from the mould

4. CAST A CEILING ROSE

OBJECTIVE

To prepare a flexible or rigid mould and produce a ceiling rose ready to fix to a ceiling.

INTRODUCTION

You can use either a metal mould or a flexible PVC mould for this task. In both cases, the procedure is very similar to casting a plain face slab. The main difference between using a flexible and rigid mould is in the preparation, as a PVC mould does not require shellac and grease. However, reinforcing materials are still required, along with the same mixing procedure.

TOOLS AND EQUIPMENT

Small tools	Running rules
Drag or busk	Hessian
Joint rule	Shellac
Splash brush	Grease
Mixing bowls	Timber
Casting plaster	

PPE

Ensure you select PPE appropriate to the job and site conditions where you are working. Refer to the PPE section of Chapter 1.

PRACTICAL TIP

Soak the laths overnight before use.

Figure 8.74 A rubber mould for a ceiling rose

STEP 1   Prepare the reverse mould by washing it with cold water, making sure all the areas are free from plaster. If you are using a rigid mould, apply grease to all parts of the inside.

Figure 8.75 Applying grease to the rigid mould

STEP 2   Next cut the facing and backing canvas. Make sure the canvas overlaps the edges of the mould by 50 mm.

STEP 3   Cut a longer piece of wooden lath to use when you need to rule the full length of the moulding, Cut shorter pieces of lath to use at the thickest points of the mould. Leave the laths to soak in water until you use them.

STEP 4   Prepare the firstings and seconds. Remember the seconds always have retarder added and so should be mixed first and leave them to soak to be used later.

STEP 5   Now mix the firstings to the consistency of double cream.

STEP 6 Pour a small amount of the firstings into the mould and brush it over the whole area.

Figure 8.76 Brushing over the firstings

STEP 7 Holding the bristles of your splash brush, splash a layer of plaster up to 3mm thick all over the reverse mould.

STEP 8 Position the facing canvas and, using your splash brush, soak this canvas with a layer of seconds.

Figure 8.77 Positioning the canvas

STEP 9 Position the backing canvas and cover it with seconds. Make sure you turn back the canvases so they do not go beyond the strike-off points.

Figure 8.78 Preparing the backing canvas

Figure 8.79 Covering the canvas with plaster

STEP 10 Now place the shorter pieces of lath at the thickest points of the mould – this is usually the centre of a ceiling rose mould. Cover it with seconds.

STEP 11 Splash a final layer of plaster over the cast to give it extra strength and thickness. Rule it off with the timber rule using the strike-off points as the edge of the moulding. Clean off the strike-off points with a joint rule or trowel.

Figure 8.80 Cleaning off the strike-off points

STEP 12 Let the cast set for approximately 30 minutes then carefully lift it from the mould. Clean the back of the ceiling rose with a busk or a joint rule, ensuring it is flat ready for fixing.

Figure 8.81 The finished ceiling rose

TEST YOURSELF

1. What is a cornice?

   a. The top part of a horizontal moulding where the wall meets the ceiling

   b. A circular mould attached to the ceiling

   c. A decorative moulding that covers a ceiling beam

   d. The moulding that forms a door frame

2. Which of these is NOT a way of getting information about the profile of the item you want to copy?

   a. Copying it from a photograph

   b. Referring to a drawing

   c. Referring to the manufacturer's data sheet

   d. Taking a squeeze

3. Why is ordinary wall or ceiling plaster not suitable for casting?

   a. It dries too quickly

   b. It will not produce enough detail

   c. It is the wrong colour

   d. It is too expensive

4. You are most likely to use a splash brush for applying plaster to which sort of mould?

   a. Flood mould

   b. Reverse mould

   c. Flexible rubber mould

   d. Running mould

5. What materials do you normally use to make a running mould?

   a. Plasterboard and aluminium

   b. PVC

   c. Cold pour compound

   d. Timber, zinc and nails

6. What is glue size?

   a. A type of casting plaster

   b. A release agent

   c. A retarder

   d. A pigment

7. Which of these is a type of false profile?

   a. A muffle

   b. A stock

   c. A strike-off point

   d. A gig stick

8. How are seconds different to firstings?

   a. They are mixed using a different grade of plaster

   b. They are mixed with less water

   c. They contain retarder

   d. There is no difference

9. What should you always do before using laths?

   a. Dry them out completely

   b. Coat them in shellac

   c. Sandpaper them until they are completely smooth

   d. Soak them in water

10. What do you use to dilute shellac?

    a. Water

    b. Methylated spirit

    c. Retarder

    d. You should never dilute it

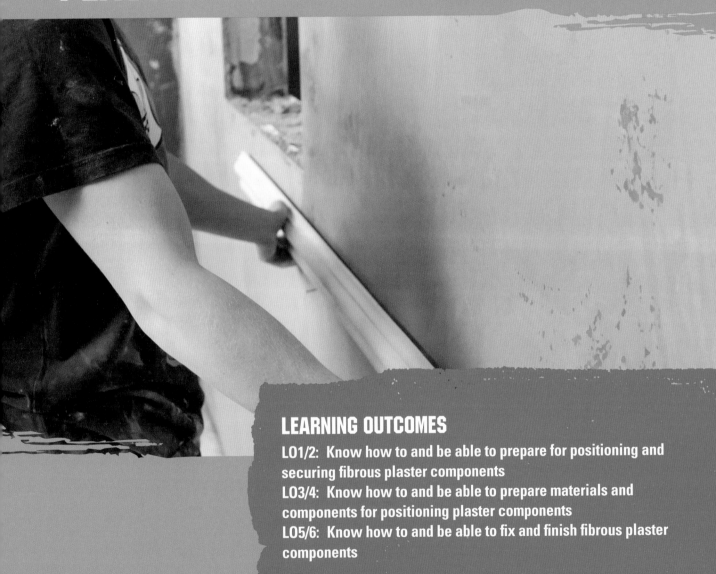

Chapter 9

Unit CSA L2Occ58
POSITION AND SECURE FIBROUS PLASTER COMPONENTS

LEARNING OUTCOMES

LO1/2: Know how to and be able to prepare for positioning and securing fibrous plaster components

LO3/4: Know how to and be able to prepare materials and components for positioning plaster components

LO5/6: Know how to and be able to fix and finish fibrous plaster components

INTRODUCTION

The aims of this chapter are to:

* show you how to make backgrounds ready for fixing plaster components

* explain how to set out and cut plaster components

* explain how to fix plaster components using different methods.

In Chapter 8, we looked at ways of casting mouldings for features such as cornices and panels. In this chapter, we will look at how these mouldings can be fixed in position, both when you are repairing existing features, and when you are replacing them or putting in new ones.

As a plasterer, you may be required to fix the mouldings even if you haven't made them yourself. They will probably have been produced in a workshop offsite and brought to the site for positioning. As you saw in the previous chapter, it takes skill and patience to produce plaster casts, especially if they are designed to match the profile of an existing feature, so you must always treat the casts with care.

How you fix the cast in position depends on:

* its size

* its shape

* its weight

* where it is going to be fixed

* the background you are going to fix it to

* the position of any reinforcement within the cast

* any particular requirements in the specification, for example whether a particular type of fixing is required to maintain the character of a heritage building.

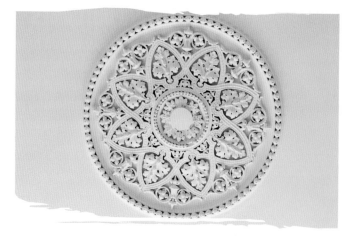

Figure 9.1 You may be required to repair heritage plasterwork on site

PREPARING FOR POSITIONING AND SECURING FIBROUS PLASTER COMPONENTS

Risk assessments and hazards

As usual, make sure a risk assessment is in place before you start work. It should include things like the following:

* **Manual handling:** Long lengths of cornice are heavy, making them difficult to carry and hold above your head. Even a manufactured cornice, with a gypsum core, weighs about 5 kg over a 3,000 mm length. You could easily strain your back and arms when trying to hold it in place, and there is also the risk of it falling on your head. Where possible, ask someone to help you. Your site or company may have approved methods of holding plaster components in place while you are fixing them.

* **Working at height:** It is likely that you will need to reach high up to fit plaster components like ceiling roses, plain-faced slabs and cornices. Use your site's preferred method of working at height, for example stilts or a platform. Ladders are not really suitable for this type of work.

* **Protecting passers by:** Make sure signs and barriers are in place to stop people entering your work area and risking injury. If you are working in someone's home, make sure they are aware of what you are doing and ask them not to enter the room until you have finished and cleared up.

* **Using power tools:** You may need to use an electric drill to fix plaster components in place. You should only use power tools after you have been trained and your supervisor says you are competent to do so. Remember to use the correct voltage for the site.

* **Dust suppression**: Cutting, fitting and sanding plaster casts creates dust. Follow the approved procedures to minimise the risk to your own health and that of people nearby.

* **PPE**: As always, wear the PPE that is appropriate to the job you are doing, and in line with your site and company rules. Due to the risk of head injuries, you may be required to wear a hard hat for this type of work when you might not usually wear one when you are plastering a wall.

Refer to Chapter 1 for more information about risk assessments and health and safety.

Information sources used when fixing fibrous plasterwork

As always, before you start work, double-check against the architect's specification and drawings that you know exactly what is required. Take particular note of any special instructions or warnings given by the person who made the mould, as they are most familiar with it.

Make sure you know which moulding will go where – they should be clearly labelled. If you are not sure, ask.

You might also be required to remove some or all of an existing moulding, for example if it is damaged. Again, make sure you are very clear about which moulding needs to be removed and how much of it has to be taken off. If you have to remove a part of it, make sure your measurements are correct, as any replacement feature will have been run off to the exact size required. If you remove too much, the moulding will not fit.

When you are setting out, you might also need to establish datum points and centre lines when locating the feature – you will use the same methods as those shown in Chapter 7.

Calculating the quantities of the materials you need

Before you start, you need to know the following:

* whether the quantity of the casts you have been given matches the specification – for example, whether you have a sufficient length of cornice to go round the room. You will need to measure the room and the casts, and check these against the specification.

* how many mechanical fixings you will need. If you are using adhesive and fixings, you will normally fix them about 300 mm apart, or 600 mm apart for larger casts; however, this assumes that there are sufficient ceiling joists or other strong backgrounds to fit them to. If you are using the hidden fixings method (see page 262) you will drive nails or screws through the laths, so you need as many fixings as laths in the cast.

* how much adhesive you need to mix. This will depend on the size of cornice and the type and quality of wall being adhered to. A high suction background will absorb much of the adhesive if you haven't prepared the wall properly, for example by coating it with PVA.

PRACTICAL TIP

Make sure you know which way up the moulding needs to be before you fix it in place as removing it could be costly if it is damaged.

PRACTICAL TIP

Don't forget to add 10 per cent for wastage – but you should try to minimise waste when fixing the components.

PREPARING MATERIALS AND COMPONENTS FOR POSITIONING AND SECURING FIBROUS PLASTER COMPONENTS

Types of plain and decorative mouldings

Table 8.2 in Chapter 8 shows different types of mouldings. Figure 9.2 shows these mouldings in place in a room. It is unlikely that one room will contain all these types of moulding but you can see where they are positioned in relation to each other.

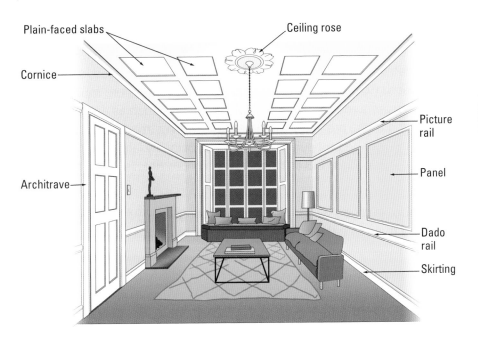

Figure 9.2 Examples of different mouldings

The effects of incorrect preparation

Plaster casts should be stored and transported as described in Chapter 8, on page 241, to ensure they are not damaged before they are fixed.

If you do not spend enough time preparing the background and plaster components before fixing them in place, it is more likely that mistakes will be made. For example, you might find that, when you are fixing the cast, the finish of the moulding may not be of high enough quality, or it may not be attached properly and can fall off.

As well as the risk of injury, problems like these are likely to be costly in terms of both time and money, and you do not want to be responsible for a project running late or going over budget.

Why the properties of fibrous plaster affect how you fix the cast

You need to know what sort of plaster the component you are fixing is made from, as this will influence:

* its weight

* its strength

* the type and position of any reinforcements, such as hessian

* how you prepare the background

* how you fix it to the wall or ceiling.

As we saw in Chapter 8, the plaster used for making casts is different from standard plaster as it is not usually gypsum-based. Most plaster components are made from plaster of Paris, because it is finer and shows details more clearly, and is soft enough to be carved, sanded and shaped. However, components that need to be more robust will be made from a much stronger grade of plaster, such as autoclaved plaster, and these are likely to be heavier. Moulds made from plaster of Paris will not be particularly heavy, but obviously larger pieces will be heavier. See Table 8.3 on page 225 for more details about types of casting plaster.

Remember that you will be working against gravity when fixing casts – you need to ensure that they are securely positioned and do not fall off the wall or ceiling.

Checking the background

As well as considering the moulding itself, it is just as important to look at the background it will be fixed to. There is no point in fixing the moulding if the background is:

* damp

* too weak to take the weight of the moulding

* high suction

* flexible

* not flat

* covered with loose material

* dirty or greasy.

Table 9.1 suggests ways in which these problems can be addressed.

PRACTICAL TIP

The background should be plumb and flat to within 3 mm over a 1.8 m length.

Background problem	Suggested next step
Damp surface	Treat cause of damp and apply a moisture-resistant coating.
Weak surface	Tap it with a hammer and listen for a hollow sound, which indicates that the plaster surface has not bonded to its background. Raise the issue with your supervisor, client or architect to find out whether the plaster component can be sited elsewhere and, if not, how the background can be strengthened.
High suction surface	Apply PVA to reduce suction and score or rough up the wall to provide a good key.
Flexible surface	Background needs to be made rigid, for example with timber reinforcements or plasterboard. Consult your supervisor or the architect.
Bumpy, pitted or broken surface	It is likely to need replastering.
Loose materials	Scrape off loose wallpaper or flakes of paint and remove lumps of old plaster. Tapping the surface with a hammer will loosen any less obvious materials that are likely to come off easily. Brush down and sand the area if necessary. If removing the loose material leaves an uneven surface, you might have to replaster it.
Dirty or greasy surface	Sponge off the dirt or grease, taking care not to over-wet the wall.

Table 9.1 Troubleshooting problems with the background

FIXING AND FINISHING FIBROUS PLASTER COMPONENTS

Types and sizes of fixings used in fibrous work

The fixings you will use to attach mouldings are similar to those used for attaching plasterboard – that is, specialist adhesives, nails, screws, bolts and plugs. See Chapter 5 for details about these. Normally you would hold the component in place using adhesive first, and fix it securely with nails or screws if it is likely to be insecure.

The length and size of the mechanical fixing depends on the size and weight of the component.

Bonding compound usually comes in 25 kg bags and is mixed with water to create an adhesive. These are often plaster-based and some types contain retarder to increase the working time.

Methods of fixing fibrous work

Fixing a cornice
The most important thing to consider when preparing to fix a cornice is its **projection** and **depth**.

PRACTICAL TIP

Screws and nails should be countersunk, and the hole filled with plaster.

DID YOU KNOW?

Mechanical fixings should be non-ferrous – that is, not contain any iron that may rust and discolour or rot the moulding.

Galvanised, zinc-plated and brass nails or screws are all suitable options.

KEY TERMS

Projection
– the measurement from the wall to the outside edge of the cornice on the ceiling.

Depth
– the measurement from the ceiling to the underside edge of the cornice on the wall.

Figure 9.3 Mixed bonding adhesive

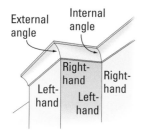

Figure 9.4 Internal and external angles

Setting out

To transfer the projection, on the ceiling, draw a chalk line a distance from the wall equal to the top measurement of the cornice.

To transfer the depth, on the wall, mark a chalk line equal to the bottom measurement of the cornice.

When you measure the amount of cornice needed for a room, you need to add the projection required for the external **mitres** (also called angles or corners) like chimney breasts or boxing.

You also need to consider where you will join lengths of cornice if they are shorter than the wall so that the join is not too obvious. For this reason, the longer the cornice, the better, although of course it is likely to be heavier and may be difficult to fix with just one person.

Measure the lengths of the chalk line around the room and check you have enough cornice, allowing for a cornice-width of overlap at external corners. With a craft/trimming knife, lightly score the area where the coving will be fixed.

Fix a nail about every 600 mm along the chalk line as a temporary support. The cornice will rest on these nails while the adhesive sets. You'll remove these nails once the cornice is secure.

To mark an internal corner: push a length of cornice into a corner and mark on the top edge where the coving crosses the line on the ceiling. Then mark the next piece of coving on the adjacent wall in the same way.

To mark an external corner: hold a length of cornice to the wall and mark on its lower edge the end of the wall. Keep the cornice where it is and mark where the top edge intersects the ceiling line. Then do the same for the cornice on the other side of the corner.

Cutting the cornice

Although it is possible that you will be required to simply fix in place the cornice you have been given, it is more likely that you will need to cut it to size or cut one or more mitres so that it fits around corners and angles in the room. You should also cut straight lengths of cornice to a mitre, to avoid a straight joint where they meet. Using a mitre joint rather than a straight cut will make the joint easier to fill and less visible when completed.

With a straight edge and a pencil:

* for an internal corner, draw a diagonal line from the marked point on the top edge to the lower corner

* for an external corner, draw a line between the two marked points on the top and lower edges and cut along the line.

You should cut the cornice upside down with the wall edge uppermost, flat against the side, and the ceiling edge flat against the base. The face of the cornice should be towards you.

Cut the corner angles with a fine-toothed saw held upright.

Although you can support the cornice on wooden squares attached to the cutting bench, many people prefer to use a mitre box (or block) because, as long as it is the right size, it stops the cornice from moving. The projection should be at a 45° angle – to do this, you might need to fix small battens to the mitre box, away from the ceiling projection.

Figure 9.6 A mitre box

Figure 9.5 Plan of a mitre box

If the wall turns left, place the mitre box to the left of the cornice. Place it on the right if the wall turns to the right.

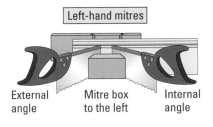

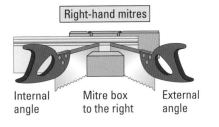

Figure 9.7 Positioning the mitre box

You can cut the mitres to allow a 5 mm gap, to allow for movement; this can be filled with casting plaster when the cornice is in place.

Check the join by holding the two pieces to the corner.

Attaching the cornice using adhesive and mechanical fixings

Before you start, key the back of the moulding and soak it in water to reduce its porosity.

You could also key the wall to increase suction.

Where possible, begin to fix the moulding in an internal corner of the longest wall that faces the doorway.

Figure 9.8 Keying the wall before fixing a cornice

Mix the adhesive (or casting plaster for smaller, light mouldings) and apply it to the entire length of the cornice, on the top and bottom edges where they will be in contact with the wall and ceiling. Rest the cornice on top of the nails and firmly press it upwards and against the wall.

When the adhesive starts to grip, scrape off any excess with a knife and then brush or sponge it off to get a clean finish.

If the cornice is particularly heavy or long, also fix it to the wall and ceiling joists with screws, 300 mm apart, or 600 mm apart for really big casts.

You may need to make vertical joints and screw holes good by filling them with a little casting plaster, mixed to the consistency of single cream.

The cornice will be ready to paint after about 24 hours in normal conditions.

Attaching the cornice using hidden mechanical fixings

This method is more difficult than using adhesive, especially if you are working on your own, but it may be preferred by the architect or client.

If possible, find someone to help you so that one person can hold the cornice in position while the other person drills a small leader (pilot) hole into the cast's laths at the strike offs through to the fixing points on the wall and ceiling. This allows a mechanical fixing to go through the hole without it showing on the face of the cornice.

Take the cornice away and drill the holes in the wall and ceiling to the required size and insert a wall plug. Put the cornice back, line up the holes and insert the fixings below the surface, making sure that the strike offs are flush with the wall and ceiling.

Fixing plain-faced slabs
Setting out

You would normally attach plain-faced slabs either to a timber ceiling or as part of a suspended ceiling.

Set out the centre line and mark in pencil or chalk where each slab will go. You can fix a string line to establish the position of the first row of slabs.

Identify the position of the timber ceiling joists and drill a pilot hole into one for the first slab.

To ensure strength, you will need to fix the slab through the lath or laths inside it, so it is helpful to drill pilot holes before lifting the slab to the ceiling.

Types of joint

Plain-faced slabs need to be joined to form an expanse of flat surface, for example in a floating ceiling. Three types of joint are suitable for this, all with a recess cast into them. This enables a wad (canvas soaked in plaster) to be inserted between the slabs after they have been fixed, which can be levelled off to hide the join. You should refer to the specification to find out which type of joint is required. If none is specified, you will need to use your judgement. Rebated joints are the

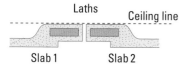

Figure 9.9 Rebated joint

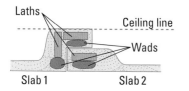

Figure 9.10 Square joint

most commonly used as they are the easiest to screw together, and allow for expansion. See Figs 9.9, 9.10 and 9.11.

Fixing the slabs to a timber ceiling

You should not have to cut the slabs to size as they should have been made to fit. Make sure the back of each slab is flat, so that it attaches flush to the ceiling. Test that each slab is flat before starting and, if required, use a rasp to smooth away any lumps or uneven edges.

Use a mechanical fixing to attach the first slab to the ceiling joist. Then position the second slab on the line and line it up with the first slab using a straight edge and wedges across the joint. Use the first two slabs as a guide for the remaining ones.

Fixing the slabs to a suspending ceiling

Plain-faced slabs can be also fixed to a suspended ceiling with the wire and wad system. You will know from Chapter 8 that wads are plaster-soaked canvas strips on the back of the cast, which can have a wire embedded in it.

Set out the centre lines and mark the fixing points for each slab. A large area of plain-faced slabs can be lined through using a builder's line.

Drill diagonal holes on each side of the lath to line up with the fixing points on the ceiling. Holding the slab in position, pass a wire through it. The slab can be levelled up or down by lengthening or shortening the wire. Once the slab is positioned, insert the wad behind the cast. Then repeat for the rest of the slabs in the row, and reposition the builder's line for the next row.

Once all the slabs are in place, use a trowel to push more wads along the length of the rebate joint to cover the gap, ensuring that none of it pokes out. Cover the area, and any other holes, with casting plaster (mixed with PVA if you need to control suction), ruling it flat and cleaning it with a tool brush.

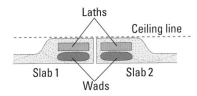

Figure 9.11 Lapped joint

PRACTICAL TIP

You will need to support the slabs with a prop or similar equipment while you are fixing them.

PRACTICAL TIP

Make sure the electrical cables are not live.

PRACTICAL TIP

Drawing two diagonal lines from corner to corner will help to locate the central point – as long as the room has straight walls.

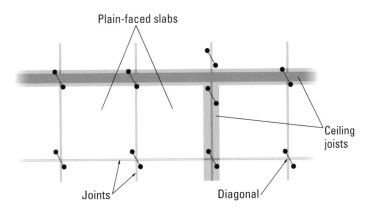

Figure 9.12 Positioning slabs on a suspended ceiling

Figure 9.13 Fixing a ceiling centre

Figure 9.14 A pilot hole

Fixing a ceiling centre

This method is for fixing a ceiling centre (rose) with a central light fitting. Do not drill a central hole in the ceiling centre moulding if it is not going to host a light fitting.

You should not handle electrical cables unless you have been trained to do so. The wires may have been prepared by the site electrician, but your job is only to feed them through the ceiling rose.

Setting out

Trace the circumference of the ceiling centre from a central point.

Drill a hole in the ceiling at the central point of where the ceiling centre will be positioned. Drill pilot holes around the perimeter of the moulding if you are going to use mechanical fixings.

Dampen the back of the ceiling centre to reduce suction.

Attaching the ceiling centre using adhesive and mechanical fixings

Apply a 10 mm line of adhesive around the perimeter of the ceiling centre, and add a little more adhesive towards the middle. Push it onto the ceiling, lining up the hole in the moulding with the hole in the ceiling by pushing a screwdriver through both holes.

As with other types of moulding, you can fix screws around the perimeter for added strength. Ensure the screw is countersunk and filled in with adhesive or plaster of Paris, using a small tool and busk.

To keep the ceiling centre in place and to maintain pressure while the adhesive dries, support it with a prop until the adhesive has dried.

Clean off excess adhesive and brush or sponge to a clean finish.

Pre-drilling pilot holes

Drilling a pilot hole, smaller than the nail or screw that will go in it, helps to prevent the nail or screw from splitting the wooden joists and cracking the plaster.

If you do not drill a pilot hole, the screw or nail may act as a wedge, generating outward pressure that can split the materials it has been driven into. A pilot hole reduces this pressure and makes the nail or screw less likely to fall out or break.

It also makes the cast much easier to fix, as you do not need to drill the holes at the same time as holding it in place.

You can enlarge the pilot hole with a busk or small tool.

Supporting heavy mouldings during the fixing process

Mouldings can be heavy and you could injure yourself or damage the moulding if you try to hold it above your head at the same time as fixing it. There are several ways of supporting the moulding:

* Work with a partner – one of you can hold the moulding while the other fixes it. Ensure that the other person is able to carry the weight without injury or discomfort.

* Using fixing blocks – temporarily nail wooden blocks to the wall to support the cornice. Obviously, a disadvantage is that this makes holes in the wall.

* Fixing nails in the wall – this is suitable for lighter components but, again, leaves holes in the wall.

* Use a prop – some plasterboard lifting tools could be suitable (see page 160) but take care that they do not damage the moulding.

Clearing up your place of work

When you have finished the job, ensure that:

* the work area is cleared and materials and adhesives are disposed of, reused or recycled in accordance with your site procedures, legislation, regulations, codes of practice and the job specification

* hazardous material is identified for separate disposal

* you clean all your tools and equipment

* tools and equipment are checked, maintained and stored in accordance with manufacturer recommendations and your site or company procedures.

CASE STUDY

South Tyneside Homes

South Tyneside Council's Housing Company

It's important to enjoy what you do

Ed Goodman at South Tyneside Homes has been in the trade for almost 30 years and has some good advice for young people.

'It's great if you can get an apprenticeship in a trade because there will be a skills gap in the future and it will be harder to get tradespeople who know what they're doing. That said, it will also help if you enjoy what you're doing – you need to want to do it right from the beginning – not because someone else wants you to do it.

I'd also tell young tradies not to rush anything – when you're learning, you're learning. Speed comes with time and experience. It's better to get it right.

You'll probably be working with people who've been in the trade for a lot longer, so keep your ears open and pick up their advice. There are lots of ways of doing things and sometimes there can be conflicting ideas. So try out different techniques and you will figure out which is best for you – though usually what you're taught in college will be the correct way, so listen to your lecturers too.'

1. FIX A CEILING CENTRE (ROSE)

OBJECTIVE

To fix a ceiling centre (also called a ceiling rose) to a prepared ceiling, ready for decoration.

INTRODUCTION

This method is for fixing a ceiling centre with a central light fitting. It is most commonly used in more traditional properties and restoration work. Use appropriate scaffold or other access equipment to work at height.

TOOLS AND EQUIPMENT

Small tool	PVA
Drag or busk	Cordless drill
Joint rule	Screws
Mixing bowls	Pencil
Casting plaster	

PPE

Ensure you select PPE appropriate to the job and site where you are working. Refer to the PPE section of Chapter 1.

STEP 1 Ensure the electricity is switched off at the mains before fixing the ceiling centre.

STEP 2 Make sure the ceiling surface is clean, sound, flat, dry and free from grease and loose particles.

STEP 3 Locate the ceiling joists at the required ceiling centre position. Where there is no convenient joist, you will need to use toggle bolts or similar fixings to secure the ceiling centre.

PRACTICAL TIP

Apply PVA to the back of the ceiling centre before applying any adhesive.

STEP 4 Position the ceiling centre and choose four fixing positions to coincide with the joists, spaced as evenly as possible. Mark them on the ceiling centre.

STEP 5 Drill countersunk holes through the front of the ceiling centre. Then drill a hole in its centre for the lighting cables to pass through.

Figure 9.15 Drilling a pilot hole

STEP 6 Check the position is correct by holding up the ceiling centre and pulling the light cables through the hole. When you are happy it is in the right place, draw a line round the edge.

STEP 7 Liberally apply adhesive to the back of the ceiling centre, thread the wiring through the centre hole and position the ceiling centre to the lines. Holding it in position, screw the ceiling centre into the ceiling joists, tightening and levelling as necessary.

PRACTICAL TIP

Be careful not to overtighten the screws. Cover the screw heads with adhesive.

STEP 8 Fill any gaps using fine casting plaster, and clean off surplus adhesive with a small tool brush.

PRACTICAL TASK

2. FIX A CORNICE, INCLUDING AN EXTERNAL AND INTERNAL ANGLE

OBJECTIVE

To fix a cornice to include an external and internal angle, ready to receive decoration.

INTRODUCTION

A fitted cornice adds elegance to a room and is usually fitted in more traditional properties and restoration work. There are many cornice designs and sizes to choose from.

In most cases two plasterers are required to carry out the fixing.

TOOLS AND EQUIPMENT

Small tools	Cordless drill
Drag or busk	Screws
Joint rule	Pencil
Mixing bowls	Mitre box
Casting plaster	Scaffold
PVA	

PPE

Ensure you select PPE appropriate to the job and site conditions where you are working. Refer to the PPE section of Chapter 1.

STEP 1 Set up a safe and stable working platform using two pairs of steps or hop-ups with a suitable scaffold board across them. Be sure that the board is properly supported and strong enough to take your weight, and that of the person working with you. Planning your access equipment will enable you to concentrate on fixing the cornice, rather than having to keep moving your steps.

PRACTICAL TIP

It is always best to work from a platform rather than simply on a stepladder.

STEP 2 Transfer the projection of the cornice onto the ceiling by drawing a chalk line a distance from the wall equal to the top measurement of the cornice.

STEP 3 Then transfer the depth of the cornice on the wall by marking a chalk line equal to the bottom measurement of the cornice.

PRACTICAL TIP

Cut off a 100 mm length of cornice to use as a template. Use this piece to mark the top and bottom edges of the cornice on the walls and ceiling all around the room. Make the marks at regular intervals.

STEP 4 Prepare the background by removing any wallpaper or loose paint and plaster from wall. With a trimming knife, make criss-cross scratches between the guidelines to provide an extra key for the adhesive.

STEP 5 Measure the perimeter of the room to calculate the amount of cornice required. Remember to allow extra for external angles.

STEP 6 If the cornice is heavy and you require secondary mechanical fixings, find the ceiling joists and make their position on the ceiling.

STEP 7 Fix clout nails about every 600 mm along the chalk line. These will help to support the cornice as you fix it in place and while the adhesive sets.

STEP 8 For the internal angle, place the cornice up to the ceiling and mark a line on the top edge where the cornice crosses the line on the ceiling. Repeat for the return piece of cornice.

STEP 9 Now place the cornice in the mitre box. Making sure it's at a 45° angle, lay the cornice upside down with the wall edge uppermost, flat against the side, and the ceiling edge flat against the base. The face of the cornice should be towards you.

STEP 10 Using a fine saw, cut through the line. Allow 5 mm for movement. Keep any offcuts of cornice as they may come in handy for short returns.

Figure 9.16 Cutting the cornice in the mitre box

PRACTICAL TIP

Take your time with cutting the mitres. The neater they are, the less work you will have later on.

STEP 11 Dry-test the cornice and the mitres to check they are in line. Small gaps can be easily filled after fixing.

PRACTICAL TIP

To improve suction before applying adhesive to the back of the cornice, or any moulding, try applying PVA, dampening it with water or scratching the surface with a trimming knife to form a key.

STEP 12 Apply an even amount of adhesive to the back of the cornice and offer it up to the ceiling. Rest it on the support nails and firmly press it into position, lining it up with the guidelines.

Figure 9.17 Applying adhesive to the back of the cornice

STEP 13 Repeat for the other lengths of cornice, and then clean off excess adhesive from the cornice with a small brush. Remove the temporary nails.

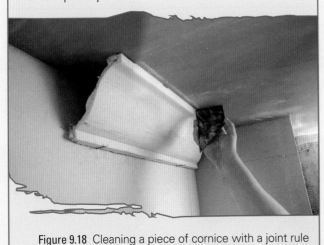

Figure 9.18 Cleaning a piece of cornice with a joint rule

STEP 14 Using fine casting plaster, fill in the gaps at the angles. Use a joint rule to follow the lines of the cornice and make good by applying small amounts of water to smooth it off ready for decoration.

PRACTICAL TASK

3. FIX A DADO PANEL MOULDING TO INCLUDE AN EXTERNAL ANGLE

OBJECTIVE

To fix a dado moulding around a room, to include an external and internal angle, ready to receive decoration.

INTRODUCTION

Although dados are traditionally used as wall protection from furniture, these days they are more often used to give a two-toned design approach, with the upper area being either of a different colour or style.

TOOLS AND EQUIPMENT

Small tool	Cordless drill
Drag or busk	Screws
Joint rule	Pencil
Mixing bowls	Mitre box
Casting plaster	Scaffold
PVA	

PPE

Ensure you select PPE appropriate to the job and site conditions where you are working. Refer to the PPE section of Chapter 1.

STEP 1 First you need to decide on the height of the dado moulding. They are normally positioned between 1 m and 1.2 m up from the floor. As a general rule, the higher the ceiling, the higher position of the dado rail.

STEP 2 Mark out the position of the dado moulding using a level and draw guidelines on the wall.

Figure 9.19 Marking out the position of the dado

STEP 3 Remove any wallpaper or loose paint and plaster from the area where you are going to fix the dado. Make criss-cross scratches between the guidelines with a trimming knife to provide an extra key for the adhesive.

Figure 9.20 Scoring the wall to provide a key

STEP 4 Measure the perimeter of the room (the length of each wall) to calculate how many lengths of dado moulding you will need. This will depend on the lengths of dado.

PRACTICAL TIP

Try to avoid too many straight joints by using longer lengths of panel moulding.

STEP 5 Fix clout nails every 600 mm along the guideline to help support the dado as you fix it in place and while the adhesive sets.

STEP 6 Mark the length of the first piece of dado and place it in the mitre box at a 90° angle. Carefully cut it with a fine saw, allowing 5 mm for movement.

STEP 7 Place the dado moulding on the wall and check that the mitre fits and is in line with the return.

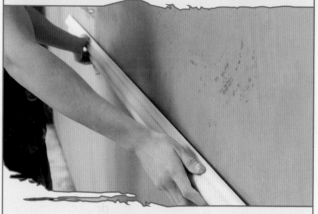

Figure 9.21 Checking the position of the dado

STEP 8 Apply an even amount of adhesive to the back of the dado moulding and place it on the wall, resting on the support nails, and firmly press it into position against the guidelines.

STEP 9 Repeat for the other lengths of dado, and then clean off any excess adhesive with a small brush. Remove the temporary nails.

STEP 10 Use fine casting plaster, fill in any gaps at the angles. With a joint rule follow the lines of the dado and make good, applying small amounts of water to smooth it off ready to receive decoration.

Figure 9.22 Making good with a joint rule

TEST YOURSELF

1. Where would you find a cornice?

    a. Around a door frame

    b. Halfway up a wall

    c. At the join between the wall and the ceiling

    d. In the middle of the ceiling

2. Which of these is NOT a joint used for plain-faced slabs?

    a. Rebated

    b. Lapped

    c. Stopped

    d. Square

3. If you hear a hollow sound when you tap a wall with a hammer, what does it indicate?

    a. That the plaster has not bonded to its background

    b. That the background is made of brick

    c. That the background can be fixed with plaster components without further preparation

    d. That suction is too high

4. What is the accepted tolerance for a flat wall or ceiling?

    a. 3 mm over a 3 m length

    b. 3 mm over a 1.8 m length

    c. 3 mm over a 2.3 m length

    d. 3 mm over a 1 m length

5. How do you measure the depth of a cornice?

    a. From the wall to the outside edge of the cornice on the ceiling

    b. From the wall to the outside edge of the cornice on the wall

    c. From the striking point to the centre

    d. From the ceiling to the underside edge of the cornice on the wall

6. How should you position a cornice in a mitre box?

    a. With the ceiling edge against the base and the face towards you

    b. With the wall edge against the base and the face towards you

    c. With the ceiling edge against the base and the face away from you

    d. With the wall edge against the base and the face away from you

7. What is a pilot hole?

    a. A hole that marks the datum line

    b. A small hole drilled before the mechanical fixing is inserted

    c. The hole in the middle of a ceiling centre when it is fixed

    d. A large hole for passing wire through when fixing slabs

8. What is the wire and wad system used for?

    a. Fixing ceiling centres

    b. Fixing a cornice to the top of a wall

    c. Fixing plain-faced slabs to make a suspended ceiling

    d. Cleaning up a plaster component when it is in position

9. Why should mechanical fixings for plaster components be non-ferrous?

    a. Because ferrous fixings are not strong enough

    b. Because ferrous fixings rust

    c. Because ferrous fixings are not long enough

    d. Because ferrous fixings are no longer legal

10. Which of these is NOT a suitable method for supporting heavy mouldings while you are fixing them?

    a. Using fixing blocks

    b. Hammering temporary nails into the wall

    c. Holding the moulding in one hand while you drill with the other

    d. Using a lifting tool